Reprints of Economic Classics

F. Y. EDGEWORTH'S

CONTRIBUTIONS TO

MATHEMATICAL STATISTICS

F. Y. EDGEWORTH'S

CONTRIBUTIONS TO

MATHEMATICAL STATISTICS

BY

A. L. BOWLEY

[1928]

AUGUSTUS M. KELLEY • PUBLISHERS
CLIFTON 1972

First Edition 1928

(London: Royal Statistical Society, *Adelphi Terrace, W. C. 2*, 1928)

Reprinted 1972 by

Augustus M. Kelley Publishers

REPRINTS OF ECONOMIC CLASSICS

Clifton New Jersey 07012

I S B N 0 678 00889 2

L C N 68-24161

PRINTED IN THE UNITED STATES OF AMERICA

by SENTRY PRESS, NEW YORK, N. Y. 10013

F. Y. EDGEWORTH'S

CONTRIBUTIONS TO

MATHEMATICAL STATISTICS

BY

A. L. BOWLEY, Sc.D., F.B.A.

LONDON

ROYAL STATISTICAL SOCIETY

9, ADELPHI TERRACE, W.C. 2

1928

NOTE

I am indebted to my colleagues at the London School of Economics for valuable assistance in the preparation of this monograph, especially to Mr. F. Brown for preparation of the typescript; to Dr. E. C. Rhodes for a critical reading of the whole exposition; and to Miss F. J. Nicholas for the compilation of the table on pp. 123–8.

A. L. B.

CONTENTS

F. Y. EDGEWORTH'S CONTRIBUTIONS TO MATHEMATICAL STATISTICS

[The numbers shown in the text thus (1) *refer to the list of publications in the Bibliography, pp.* 129 *seq. below.]*

CHAPTER I

INTRODUCTION

A YEAR before Professor F. Y. Edgeworth's death in February, 1926, the Royal Economic Society published three volumes of his studies in economics,* selected, edited and revised by himself. It is well known that his interest in mathematical statistics was as great as in economics, and it was recognised that it would have been desirable to make his statistical studies equally accessible; but I think that he would have hesitated to republish the papers as originally written, for the reason that in each subject the analysis was extended in each successive paper and much of his earlier work was superseded by the later.

It may be doubted whether his work in statistics was as well known among statisticians as was his work in economics by economists. There are several reasons for this difference. Difficult and subtle as were his economic studies, a great part was written in terms familiar to all economists, and even his mathematical economics was within the range of those economists who were familiar with mathematical analysis. His contributions to economic theory were recognised and his critical power was well appreciated by serious economists. On the other hand, his approach to statistics was from the side of the philosophy of probability; almost his earliest published writing was a critique of Venn's *Logic of Chance* in 1883 (5). At that date very little interest was taken in any but objective statistics, except in relation to the Registrar-General's returns and in actuarial

* *Papers Relating to Political Economy,* Macmillan, 1925.

science. The path-breaking paper, read at the Jubilee meeting of the Statistical Society in 1885 (9), appears to have covered ground quite unfamiliar to his audience, and to most of those who subsequently studied it the formulæ and methods were new. When the application of mathematical methods to statistics developed in this country in the early 'nineties, it followed Galton's conceptions, appropriate primarily to biology and anthropometry, rather than Quetelet's far-reaching applications of the normal curve of error.* Edgeworth, on the other hand, continued to devote himself to the conception of pure probability, while the problems he attacked appeared to be used rather as illustrations of theory than as having practical importance, though this appearance was often illusory. Exception must be made of his work on index numbers (19), which is probably regarded as his most important practical work; based though it was on the theory of probability and particularly on the law of error, it nevertheless appealed strongly to Giffen and other non-mathematical statisticians. In subsequent years he continued in paper after paper to develop the conceptions underlying the law of error, always with the intention of finding how far there was a universal law, resulting from a complex of causes, whose action could be traced in objective statistics—a philosophical conception rather than a practical analysis. Neither philosophers nor mathematicians have been much interested in this inquiry, perhaps because the philosophers were not mathematicians, and the mathematicians who were philosophers were concerned rather with the foundations of mathematics, which did not involve questions of probability. The practical statistician was generally content to apply the method of least squares without any close inquiry as to its basis. A further impediment was Edgeworth's continual insistence on the necessity of inverse probability before measurement could be related to credibility; and this method has so far been condemned or misunderstood alike by mathematicians and statisticians that it is not to be found in the majority of mathematical and statistical text-books.

An important contributing cause to this partial neglect and ignorance of Edgeworth's contributions is to be found in the manner of the writings and the form of their publication. I have

* Edgeworth's use of the modulus instead of the "new-fangled" standard deviation illustrates how individual was his line of analysis.

found in all more than seventy books and papers (apart from reviews) issued in the forty-four years, 1883–1926; of these 31 are in the *Journal of the Royal Statistical Society*, 26 are in the *London, Edinburgh and Dublin Philosophical Magazine*, and 18 in eleven other Magazines and Transactions. The first two groups are often parallel and it is necessary to collate them to get complete proofs. Each time a problem was attacked afresh there were references to the results of former work, just as an Act of Parliament is to be interpreted by reference. The summaries now and then given (in the *Encyclopædia Britannica*, 1902 (46) and in 1911 (53), and the Presidential Address to the Royal Statistical Society in 1912 (54)), are only completely intelligible after study of the earlier work. In the individual papers the mathematical style varies from complete lucidity to very involved work set out verbally and with insufficient explanation of terms, and there are not infrequent misprints, which are sometimes corrected in later papers. The pure mathematician may find a stumbling-block in Edgeworth's use of metaphors and literary allusions, though these are of great importance in showing the scope of the adjoining analysis. In fine, there is a danger that the original and important contributions which Edgeworth made to the foundation and application of mathematical statistics are at present little known, and may be lost except to a few conscientious students. The present study is an attempt to prevent this disaster.

The task is not a light one, nor has it been easy to decide what to include and what to omit when the gist of more than seventy papers is reduced to some one hundred pages. I attempt it as a duty owed to the memory of one from whom I obtained a very great part of what knowledge of statistical theory I possess—a debt repeatedly increased from the year 1895, when he first introduced me to the older authorities, until the year of his death. From frequent conversations and correspondence I have been able to understand the objects of his work and have often obtained his interpretation of passages I found obscure. This year I have read or re-read practically the whole of the papers, and traced again the development of his thought from the first definition of probability to his last extension of the method of translation. The impression obtained is that his work is singularly consistent; the germs of the last analysis are to be found in his earliest papers. In the arrangement of subjects I have endeavoured to follow the logical sequence that was always present to his mind. But I have ventured to make a complete rearrangement of the work, following out each subject from the earliest to the latest treatment and superseding the earlier by the later, rather than repeating methods and formulæ subsequently improved or discarded. In doing this I have found it necessary to re-write his analysis rather than to quote it, for in giving his final statement it was necessary to incorporate passages only given by reference to former papers. In this re-writing I have had in many cases to choose new symbols, since he varied them from paper to paper and the same letter was used with different meanings. Once it was decided to re-digest the analysis rather than repeat it in the original words, it was evident that the niceties and exuberances of the style must be sacrificed and a colourless treatment be substituted. I can only hope that I have preserved the essence of the arguments and not unconsciously substituted my own views for the author's. After all, many of the important papers are readily accessible, and it will be easy for readers to compare my rendering with the original.

It has not been possible to include an abstract of all the papers, a list of which will be found in the bibliography. Those omitted are either brief and unimportant, or they are in illustration of the methods which are described, or they contain the solutions of special problems. In such cases a very brief summary is given in the bibliography. The logical sequence of the whole is, I think, to be found in the following sections as they are arranged:—First comes "Probability" and "Credibility," in which the philosophic basis of the whole is laid. Secondly, "The Law of Error" and "The Method of Translation," in which the implications of the postulate of plural causation are worked out in the light of the theory of pure probability. Thirdly, "Applications to Special Problems," where, cases being taken to which the theory is definitely applicable, the use of the method in measuring variation, and in distinguishing the accidental (or random) from results of direct causation ("the elimination of chance") is illustrated in many practical statistical problems. Edgeworth held very definitely the opinion that it was not sufficient to measure the variation of a statistical result simply by the statement of a standard deviation, but that it was necessary to connect this standard deviation with a law of error, to assign the probability that it (or a multiple of it) would be exceeded, and to relate this probability to credibility by the inverse method. The modulus in his use always performs this complete function. Fourthly, a short section on " The Best Mean " is mainly devoted to an explanation of Edgeworth's championship of the median, which depends on an understanding of parts of the previous sections. There follows an account of his early contributions to the theory of correlation, so that it may be determined how far a claim for priority in its development is valid, and finally a note on his conception of the relations between the theories and methods of probability and of political economy.

CHAPTER II

PROBABILITY

EDGEWORTH'S statistical investigations are definitely confined to that region in which considerations of probability, as contrasted with direct enumeration, are involved. His starting-point is therefore the metaphysical conception of probability, and his first main problem the possibility of measuring gradations of belief. There is found a very marked consistency of ideas throughout his papers, whichever line of investigation is followed, and thus his conception of probability can be gathered from many of his articles; but his original and his final thoughts on this subject are best learnt from the two papers entitled "The Philosophy of Chance" in *Mind*, 1884 and 1922, in the first of which he considers Dr. Venn's *Logic of Chance*,* and in the second Mr. J. M. Keynes's *Probability*, published in 1921. The second paper, "A Reconsideration of Views Expressed in 1884," shows some amplification of, but no radical change in, the opinions expressed in the first. These cannot readily be summarised, and to understand completely the treatment it is necessary to read *in extenso* the earlier paper with continual reference to Dr. Venn's treatise; but it is perhaps possible to show briefly that part of the discussion which is more closely connected with mathematical statistics.

Probability relates to partial or incomplete belief. Gradations of probability correspond to gradations of belief. The calculus of probability is concerned with the estimation of degrees of probability. "That probability in general presents gradations has not been disproved. It seems usually possible to arrange probabilities relating to different cases, if *in eodem genere*, in an order of magnitude, to say that one is greater than another" ((69) *Mind*, 1922, p. 279). "Can we measure degrees of belief? I am not concerned to defend the numerical precision of the

* First edition 1867, second 1876, third 1888.

measurement. We may be content with the conception of measurement which satisfies the physicists. As Prof. Love writes: 'The capacity of numbers to answer questions of how many and how much . . . may be regarded as a secondary property derived from the more fundamental one of expressing order' " (*id.* p. 258). Numerical measurement is only possible by reference to a series (in Dr. Venn's use of the word) of experiments or experiences in which the proportion of occurrences of events tends to a definite limit, as the ratio of heads to tails when a coin is tossed again and again. Such measurements in general give the same order as that of gradations of belief ((5) *Mind*, 1884, p. 224), but not universally (*Mind*, 1922, p. 279). In the paper, " On the Probable Error of Frequency Constants " (51), we find more definitely that " the only postulate [implied in the analysis] is that the credibility of an event is measured by the frequency with which it occurs in the long run " ; but this leaves open the question whether credibility so *measured* is identical with belief as *felt*, a question which must be left to the psychologists; but " the relation between the fact of statistics and the feeling of belief is not much more obscure because the things related are not precisely measureable " (*Mind*, 1884, p. 235).

A Priori and *A Posteriori* Probability

But in order to connect the measurement of probability with belief it is necessary to know what series is appropriate. In a perfect die the (*a posteriori*) probability that 5 should come uppermost is $\frac{1}{6}$, the fraction of successes to trials approximated to in the long run. But the question remains, what is the (*a priori*) probability that the die is perfect or is loaded? It is argued that in every probability measurement there is involved an element of *a priori* probability.* There is a distinction between the treatment of inverse probability in cases, on the one hand, where the *a priori* chances of alternative schemes (urns containing different proportions of black and white balls) are known and it is asked from which urn a particular result

* Cf. (7) *Phil. Mag.*, Sept. 1884. " *A priori* Probabilities." It is shown that in determining a mean by least squares it is assumed that prior to observation all values are equally probable, and that in calculating *a posteriori* that two sets of observations result from different means, there is some assumption as to the probability that the two means should be identical.

was drawn, and, on the other hand, where the *a priori* chances are not known—which are the only cases where a problem arises. Edgeworth held that there is in very many groups of important cases sufficient knowledge of *a priori* probabilities: instances are as follows :—

(*a*) Let α, β be the unknown *a priori* chances that a coin is normal or is loaded. Suppose a head to be thrown n times in succession. Let q be the unknown probability that, if the coin be loaded, n heads would be thrown.* Then the chance that it was loaded is $\beta q \div \left(\beta q + \frac{\alpha}{2^n}\right)$, which tends to unity (certainty) as n increases, unless βq is small. Thus amplification of evidence (continued succession of heads) makes up for ignorance of the *a priori* chances ((20), pp. 4 & 6). Similarly the Rule of Succession that if n successes have been recorded the chance of the next trial being a success is $\frac{n}{n+1}$, may be defended in many cases if n is very great, but not if n is small.

(*b*) In the important statistical problem of determining the mean or constants of a frequency curve of known form from a sample of n objects, it becomes indifferent what function $\chi(x)$ is assumed for *a priori* values of the mean or constants if n is great and it can be postulated that $\chi'(x) \div \chi(x)$ is finite. (See p. 12 below.)

(*c*) There is often a general body of experience to be drawn on, such as that digits from 0 to 9 occur with equal frequency, which may be substituted for exact knowledge of *a priori* probability.

If none of these methods is appropriate, there must be an element of conjecture; in the absence of all knowledge there is no alternative to the assumption of equal *a priori* probabilities for the possible alternative causes.† But very slight knowledge is often sufficient to establish a greater or less (though not a

* It would seem more natural to suppose that q depends on n, and to write p^n as the chance of throwing n heads when the coin is loaded. Since p in this case may be taken as greater than $\frac{1}{2}$, the limit of the fraction tends to unity as n increases, as in the statement in the text.

† The method of treating as equal things between which no difference is known or discerned is not peculiar to probability. In economics, for example, it is often assumed that each of a class of persons has the same needs and the same satisfactions, if equality of distribution is advocated. (*Papers Relating to Political Economy*, Vol. II, pp. 387–428 *passim*.)

measured) probability, and this may be sufficient for judgment. The science of probability is not peculiar in involving some mixture of assumption.

The insistence on the necessity of some attention to *a priori* probability before the accuracy of statistical quantities can be measured, or at least pass from the region of measurement to the region of belief, marks the difference between the standpoint of Edgeworth and that of many other mathematical statisticians. There appears to be no variation in this view in any of the sequence of papers covering forty years.

Measurement of Precision

One of the principal problems in statistics is to measure the accuracy of averages and of the constants involved in frequency curves when they have to be determined from a finite number of observations. Edgeworth gave much attention to this problem from his earliest writings onward; in a group of papers (summarised below, pp. 98 *seq.*) he discussed the question, "What is the best mean?" and also the best value of the modulus. In 1908 (51) he investigated generally and in detail the method of determining the variation of the principal frequency constants in use.* The following pages are devoted to a summary of this paper, with reference to the preliminary ideas expounded in earlier papers. "The most recondite postulate is that the credibility of an event, such as the occurrence of an assigned deviation, is measured by the frequency with which such an event occurs in the long run" ((51), p. 384). This statement marks the relationship of the whole paper to the philosophical treatment of probability described above.

It is assumed that the functional form of a frequency group is known, but not its centre, average, modulus or any other constant which determines the particular curve. The data are the values of x, y, etc. (the measured characteristics of an object) for each of n objects selected at random and without replace-

* The use of the term "probable error" (or rather of the measurement described by the term) in this paper is justified by the fact that it "corresponds to the one definite notch on the scale of probability, the point of complete uncertainty whether an event such as the occurrence of an error in excess of the assigned limit will or will not occur" ((51), p. 382). The investigation is, however, based on the modulus, which, since the frequencies are generally normal, is a constant multiple of the probable error.

ment from such a group. The *quæsita* are the constants which determine the group. Attention is practically confined to the case where n is large, and the solution is necessarily approximate, generally correct to terms in $n^{-\frac{1}{2}}$.

The Best Mean

In the determination of the constants we are concerned (1) with their "best" values and (2) with the variation from these best values. A distinction is drawn between the "most probable value" of the *quæsitum*, that which gives the frequency group from which the observations would be expected with least improbability, and the value which gives "least detriment" (minimum disadvantage). The analysis is in general directed to the determination of the "most probable value."

As regards minimum disadvantage, when one "subjective" mean is to be put for many real things, we have Edgeworth's treatment in 1887 ((18), pp. 160–1), and elsewhere. Write $F(\epsilon)$ for the disadvantage accruing by putting on any occasion $x + \epsilon = l$ instead of a real value x. Laplace chose the function which gave the least mean error, Gauss the least mean square error, and others the least probable error. Edgeworth rejects all these as arbitrary, and postulates only that $F(0) = 0$, $F'(\epsilon)$ is positive, and $F(\epsilon) = F(-\epsilon)$. If then x is the real measure of an object and l is put for it (as an arithmetic mean is put for a number of quantities), the total detriment incurred in the long run by putting l on each occasion for the true value is

$$\int_l^{\infty} F(x - l)y \,.\, dx + \int_{-\infty}^{l} F(l - x)y \,.\, dx = E,$$

where $y = f(x)$ is the frequency curve under which the objects range. If y is symmetrical, E is minimised * when $l = 0$, that is, when the centre of symmetry of $y = f(x)$ is put for the real measure. If y is unsymmetrical, in the sense that the preponderating side of the curve is in the positive direction of x, then E is minimised when l is positive, and some value greater than the arithmetic mean gives minimum detriment. For example, the probability that given observations would arise as

* It can be shown that $\frac{dE}{dl}$ is 0 and $\frac{d^2E}{dl^2}$ positive when l is 0.

a random selection from a normal curve, whose centre is known and whose modulus is c, is $y = \pi^{-\frac{n}{2}} c^{-n} e^{-\frac{\Sigma d^2}{c^2}}$ (p. 13 below), where the d's are the deviations of observations from the centre. This curve is positively unsymmetrical, and the value of c of minimum detriment is therefore greater than the most probable value, which is $\sqrt{2\Sigma d^2/n}$. The writers who give $\sqrt{2\Sigma d^2/n-1}$ for the best value (p. 16 below) therefore move it in the right direction. Also if the median of the curve is taken instead of the greatest ordinate the change is in the right direction.

No general answer can be given to the question, Which mean is best in respect of precision? it depends *inter alia* on what is the purpose of the mean. "Is the shoemaker who makes a greater proportion of exact fits (who has a smaller probable error), but when he does make a misfit makes a terribly painful one, necessarily the most advantageous?" ((2), p. 364). When the analysis of the variation from the most probable value is completed, there is still the generally unknown element of detriment to be considered.

As argued above (p. 7), the principle of *a priori* probability is involved before we can relate our measurements to credibility. It cannot be assumed in general that all probable values of the *quæsitum* are equally probable, not only because there is usually no definite knowledge about the subject, but also because if one measurement (*e.g.*, c, the modulus) is taken as the *quæsitum* and its values were equally probable then the reciprocal or other function of that measurement (*e.g.*, the precision, $h = \frac{1}{c}$), which may be as good a measurement, would not then satisfy the condition exactly. When, however, n is great, only a minimum of knowledge is needed about the *a priori* probability to render sufficiently valid the assumption that all values of the *quæsitum* in the *relevant region* are equally probable. Thus, using the results obtained immediately below, let $P = He^{\phi(x_1 - x)}$ be the probability that an object x_1 will be drawn from a frequency group whose mean value is x. Let $\chi(x)$ be the *a priori* chance that the unknown mean value is x. Then $P_n = \chi(x) \,.\, Ke^{\Sigma\phi(x_t - x)}$ is the probability that, $x_1 \ldots x_t \ldots x_n$ having been found in a random selection, the true mean of the curve was x. (K is a constant such that the aggregate of the probabilities is unity.)

P_n is a maximum when $\dfrac{\chi'(x)}{\chi(x)} + \Sigma\dfrac{d}{dx}\phi(x_t - x) = 0$. Let $x = x_0$ be the solution of this equation. Then

$$\frac{1}{n} \cdot \frac{\chi'(x_0)}{\chi(x_0)} + \text{Mean } \frac{d}{dx}\phi(x_t - x_0) = 0.$$

In the cases discussed Mean $\dfrac{d}{dx}\phi(x_t - x_0)$ is finite, and therefore, unless $\dfrac{\chi'(x_0)}{\chi(x_0)}$ is very great, the first term has little influence in determining x_0 when n is great. What influence it has depends on the variation of $\chi(x)$ in the neighbourhood of $x = x_0$, and if $\chi(x_0)$ is, as is not uncommon, near the maximum value of $\chi(x)$, $\chi'(x)$ is necessarily small if $\chi(x)$ is continuous. Thus so far as the value of x_0 is concerned, it not uncommonly comes to the same thing as if we assumed equal *a priori* probability for values of x, which would give $\chi'(x_0) = 0$.

If the central value x_0 is not affected, neither is the standard deviation (σ) of the group of which it is a mean. For, as shown below (p. 19), $\sigma^2 = -1 \div n$ Mean $\dfrac{d^2}{dx_0{}^2}\phi(x_t - x_0)$. Hence σ is of order $\dfrac{1}{\sqrt{n}}$ if Mean $\dfrac{d^2}{dx_0{}^2}\phi(x_t - x_0)$ is finite (as is generally the case), and the variation in σ due to a small change in x_0 is unimportant; especially so, since nothing is gained by great precision in the measurement of σ, even if the question of minimum detriment could be answered. It will be found (pp. 16 *seq.*) that the same argument applies when we are dealing with several *quæsita*, including secondary constants (modulus, etc.). In the sequel the work is done without further reference to the detriment of errors or to *a priori* probability.

The Genuine Inverse Method: Particular Cases

Edgeworth describes his method as the "genuine inverse method." It consists in passing from the measurements (*observations déjà faites*) to their unknown source, and computing the probability that they would have arisen from sources variously placed. Thus in the simplest case—observations made from a normal group with known modulus—the probability that $x_1 \ldots x_t \ldots x_n$, average $\bar{x}$, would have been drawn from the group $y = \frac{1}{c\sqrt{\pi}} e^{-\frac{(x - \bar{x} + \epsilon)^2}{c^2}}$ is $P_0 e^{-\frac{\epsilon^2}{c^2/n}}$, where P_0 is the probability that they would have arisen from the curve with average $\bar{x}$ which makes the probability greatest. Then, as argued just above, the probability that the average of the curve was $\bar{x} + \epsilon$ is $\chi(\bar{x} + \epsilon) \,.\, P_0 e^{-\frac{\epsilon^2}{c^2/n}} \times \text{constant} = K e^{-\frac{\epsilon^2}{c^2/n}}$, where K is a constant, since $\chi(\bar{x} + \epsilon)$ is regarded as constant. It is then as likely as not that the unknown average was within the limits $\bar{x} \pm \cdot 477 \frac{c}{\sqrt{n}}$.

These principles are then applied to a number of cases of increasing complexity.

Case I ((51), p. 384). *Data:* observations from normal curve whose modulus is given; *quæsitum:* the centre of the curve.

Let $x_1 \ldots x_t \ldots x_n$ be measurements of n things drawn independently and at random from a frequency group whose equation is $y = \frac{1}{c\sqrt{\pi}} e^{-\frac{(x - x_t)^2}{c^2}}$, where x is the abscissa of the centre and $x - x_t$ any deviation. Then the probability that the n observations named would be drawn from the group is

$$P = (\pi c^2)^{-\frac{n}{2}} e^{-\frac{\sum_1^n (x - x_t)^2}{c^2}}.$$

The value of x which makes this a maximum is given by $\frac{\partial P}{\partial x} = 0$.

Take logarithms and then differentiate and equate to zero:

$$\frac{1}{P} \cdot \frac{\partial P}{\partial x} = -\frac{2}{c^2} \Sigma (x - x_t) = 0$$

$\therefore\ n x_0 = \Sigma x_t$, when x_0 is the required value;

$\therefore\ x_0 = \bar{x}$, where $\bar{x}$ is the average of the observations.

Write P_0 for the value of P when $x = x_0$, keeping P for the value when $x = x_0 + \epsilon$

$$P = (\pi c^2)^{-\frac{n}{2}} e^{-\frac{\sum_1^n (x_0 + \epsilon - x_t)^2}{c^2}} = P_0 e^{-\frac{n\epsilon^2}{c^2}},$$

where $$P_0 = (\pi c^2)^{-\frac{n}{2}} e^{-\frac{\Sigma (x_0 - x_t)^2}{c^2}}.$$

Therefore the chance that $x_1 \ldots x_t \ldots x_n$ would be selected from a curve whose centre was distant ϵ from x_0 is expressed by a normal curve in ϵ, with modulus $\frac{c}{\sqrt{n}}$. Conversely, from the principle developed above, the chance that the centre of the curve was at $\bar{x} + \epsilon$ is $Ke^{-\frac{n\epsilon^2}{c^2}}$, where K is chosen so as to make the integral of the curve unity. $\left(K = \frac{\sqrt{n}}{c\sqrt{\pi}}\right)$.

Case II. *Data:* observations from normal curve whose centre is given; *quæsitum:* the modulus.

Take the given centre as origin, and the equation of the curve as $y = \frac{1}{c\sqrt{\pi}}e^{-\frac{x_t^2}{c^2}}$. We have to find the value of c which satisfies $\frac{\partial P}{\partial c} = 0$, where $P = (\pi c^2)^{-\frac{n}{2}}e^{-\frac{\Sigma_1^n x_t^2}{c^2}}$.

$$\frac{1}{P} \cdot \frac{\partial P}{\partial c} = -\frac{n}{c} + \frac{2}{c^3}\Sigma x^2_t = 0.$$

Write c_0 for the solution of this equation; then $c_0^2 = \frac{2}{n}\Sigma x_t^2$.

Write P_0 for P, when $c = c_0$, P being the value for $c = c_0 + \gamma$.

Then $$\log P_0 = -\frac{n}{2}\log \pi - n \log c_0 - \frac{n}{2};$$

$$\log P = -\frac{n}{2}\log \pi - n\log c_0 - n\log\left(1 + \frac{\gamma}{c_0}\right) - \frac{n}{2}\cdot\frac{c_0^2}{(c_0+\gamma)^2}$$

$$= \log P_0 + \frac{n}{2} - n\left(\frac{\gamma}{c_0} - \frac{1}{2}\cdot\frac{\gamma^2}{c_0^2} + \frac{1}{3}\frac{\gamma^3}{c_0^3} - \ldots\right)$$

$$- \frac{n}{2}\left(1 - 2\frac{\gamma}{c_0} + 3\frac{\gamma^2}{c_0^2} - 4\frac{\gamma^3}{c_0^3} + \ldots\right)$$

$$= \log P_0 - n\frac{\gamma^2}{c_0^2} + \text{terms in } \frac{n\gamma^3}{c_0^3}.$$

Here if we write $n\frac{\gamma^2}{c_0^2} = \xi^2$, $\frac{n\gamma^3}{c_0^3} = \frac{\xi^3}{\sqrt{n}}$ and is of order $\frac{1}{\sqrt{n}}$ compared with ξ^2.

Neglecting terms of order $\frac{1}{\sqrt{n}}$ we have

$$P = P_0 e^{-\frac{n\gamma^2}{c_0^2}}.$$

Therefore the chance that $x_1 \ldots x_t \ldots x_n$ would be drawn from a curve whose modulus was $\sqrt{\frac{2}{n}\Sigma x_t^2} + \gamma$ is expressed by a normal curve in γ, with modulus $\frac{c_0}{\sqrt{n}} = \frac{1}{n}\sqrt{2\Sigma x_t^2}$. Conversely, the chance that the modulus was $c_0 + \gamma$ is $Ke^{-\frac{n\gamma^2}{c_0^2}}$, where $K = \frac{\sqrt{n}}{c_0\sqrt{\pi}}$.

Edgeworth then shows ((51) *Stat. Journ.*, 1908, p. 389) that the same result is obtained if the frequency curve was not normal but determined by the generalised law of error (see pp. 39 below) to which the normal curve is the first approximation. The terms which express the additions made by the second and further approximations involve the factor $\frac{1}{\sqrt{n}}$, and it is shown that they do not affect the equation just given where $\frac{1}{\sqrt{n}}$ is negligible in comparison with 1.

Case III. *Data*: observations from normal curve; *quæsita*: centre and modulus.

"The obvious course is to combine the two preceding solutions" (p. 393), and so find that the chance that the observations would be drawn from a normal curve with centre $\bar{x} + \epsilon$ and modulus $\sqrt{\frac{2}{n}\Sigma(x_t - x)^2} + \gamma$ is represented by the surface

$$y = Ke^{-\frac{n\epsilon^2}{c_0^2} - \frac{n\gamma^2}{c_0^2}}$$

where $c_0^2 = \frac{2}{n}\Sigma(x_t - \bar{x})^2$, and $\bar{x} = \frac{1}{n}\Sigma x_t$.

If, however, we ask what is the value of c, which makes the selection most probable on the average of all the values that x, the centre of the curve, can assume, we must proceed by another method ((51), p. 394) as follows:—

The chance that $x_1 \ldots x_t \ldots x_n$ (average $\bar{x}$) should be drawn from a curve whose centre is x and modulus c, is

$$y = (\pi c^2)^{-\frac{n}{2}} e^{-\frac{\Sigma(x - x_t)^2}{c^2}} = (\pi c^2)^{-\frac{n}{2}} e^{-\frac{\Sigma x_t^2 - n\bar{x}^2 + n(x - \bar{x})^2}{c^2}}.$$

The aggregate of the chances for all values of x * is

$$P = (\pi c^2)^{-\frac{n}{2}} e^{-\frac{\Sigma x_t^2 - n\bar{x}^2}{c^2}} \int_{-\infty}^{\infty} e^{-\frac{n(x-\bar{x})^2}{c^2}} dx$$

$$= (\pi c^2)^{-\frac{n}{2}} e^{-\frac{\Sigma x_t^2 - n\bar{x}^2}{c^2}} \cdot \frac{c\sqrt{\pi}}{\sqrt{n}}.$$

Then $\frac{\partial P}{\partial c} = 0$, if c satisfies $-\frac{n-1}{c} + \frac{2}{c^3}\left(\Sigma x_t^2 - n\bar{x}^2\right) = 0$,

i.e. if $c^2 = \frac{2\Sigma(x_t - \bar{x})^2}{n-1}$.

This is the value given by Gauss, Airey and Czuber, but Edgeworth holds that the first solution, where n is the denominator, is that appropriate to the problem as it generally arises. (See, however, p. 11 above.)

The "Genuine Inverse Method": General Solution

The four cases just analysed are particular examples. The general problem may now be considered. From a frequency surface $w = f(x, y, z \dots a, b \dots)$ whose form is given, $x, y, z \dots$ being measured characteristics of one object and $a, b \dots$ constants of the surface, n objects are taken independently and at random and are found to have the measurements $x_1, y_1, z_1 \dots x_t, y_t, z_t \dots x_n, y_n, z_n \dots$ Required to determine from these data the most probable values of the following: the averages of the x's, y's, z's, etc. (primary constants), the values of a, b, etc. (secondary constants), the moments and other functions of $x, y, z \dots$, and the probable errors of all these quantities when found.†

Replace $f(x, y, z \dots a, b \dots)$ by the form

$$He^{\psi(x'-x,\ y'-y,\ \dots\ a,\ b\ \dots)},$$

where x', y' ... are the values of the primary constants, x,

* In the following integration it is assumed that all values of x are equally probable. It can no doubt be shown as before that the introduction of a function $\chi(x)$ for the *a priori* chance of x introduces a term of an order which has negligible influence on the result.

† At this point Edgeworth assumed the analysis given by Prof. Pearson (*Phil. Trans.*, A., 191, 1898), but he used it in a new setting, and it seemed best to re-write the analysis so as to fit Cases I to IV above and obtain the formulæ used in the sequel *ab initio*. In the following pages the transition from Prof. Pearson's analysis (as I understand it) to Edgeworth's applications is gradual.

y . . . observations, and therefore $x' - x$, $y' - y$ . . . deviations, H being a constant chosen in each case to make the total of the chances equal to unity. We have to determine values for x', y' . . . a, b . . .

Let x_0, y_0, z_0 . . . a_0, b_0 . . . be the values of the primary and secondary constants in the surface from which the given observations would arise with least improbability.

Let $x_0 + \xi$, $y_0 + \eta$, . . . $a_0 + \alpha$, $b_0 + \beta$, . . . be any other values of the said constants.

The probability that the object whose measurements are x_t, y_t . . . would be drawn from the curve whose constants are these, may be written

$$P_t = He^{\Psi(x_0 + \xi - x_t,\ y_0 + \eta - y_t\ \dots\ a_0 + \alpha,\ b_0 + \beta\ \dots)}.$$

Then P, the probability that the n objects should be drawn whose measurements are x_1, y_1 . . . x_t, y_t . . . x_n, y_n,

$$= P_1 \times \dots \times P_t \times \dots \times P_n,$$

so that $\log P =$

$$n \log H + \sum_1^n \Psi(x_0 + \xi - x_t,\ y_0 + \eta - y_t \dots a_0 + \alpha,\ b_0 + \beta \dots).$$

Here x_t, y_t . . . are given, x_0, y_0 . . . a_0, b_0 . . . are to be determined so as to maximise P, and then ξ, η . . . α, β . . . define the universe from which the objects are drawn.

Write P_0 for the value of P when $0 = \xi = \eta = \dots$, and $0 = \alpha = \beta = \dots$

Then by Taylor's theorem

$$\log P = \log P_0 + \left(\xi\frac{\partial}{\partial\xi} + \eta\frac{\partial}{\partial\eta} + \dots + \alpha\frac{\partial}{\partial\alpha} + \beta\frac{\partial}{\partial\beta} + \dots\right)\sum_1^n\Psi$$
$$+ \frac{1}{2}\left(\xi\frac{\partial}{\partial\xi} + \dots + \alpha\frac{\partial}{\partial\alpha} + \dots\right)^2 \sum_1^n\Psi + \dots$$

where ξ, η, . . . α, β . . . are to be equated to zero in the differentials after differentiation.

The values of x_0, y_0 . . . a_0, b_0 . . . are to be found from the maximising equations $0 = \frac{\partial}{\partial\xi}\sum_1^n\Psi = \dots = \frac{\partial}{\partial\alpha}\sum_1^n\Psi = \dots$

and consequently the linear term in the equation vanishes.

In general the quadratic term is of order unity, ξ, η . . . α . . .

are therefore of order $\frac{1}{\sqrt{n}}$ and the further terms of the expansion are of order $n^{-\frac{1}{2}}$, n^{-1}, . . . (unless Ψ'', Ψ''', . . . become infinite) and may be neglected when n is great.

We have therefore $P = P_0 e^{\frac{1}{2}\left(\xi\frac{\partial}{\partial\xi} + \ldots + a\frac{\partial}{\partial a} + \ldots\right)^2 \overset{n}{\underset{1}{\Sigma}}\Psi}$ when n is great.

If we are dealing with only one variable,

$$P = P_0 e^{\frac{1}{2}\xi^2 \frac{\partial^2}{\partial\xi^2}\overset{n}{\underset{1}{\Sigma}}\Psi} \text{ or } P = P_0 e^{\frac{1}{2}a^2 \frac{\partial^2}{\partial a^2}\overset{n}{\underset{1}{\Sigma}}\Psi},$$

according as the variable is a primary or secondary constant.

Hence the standard deviation of ξ taken alone is $\sqrt{\dfrac{-1}{\frac{\partial^2}{\partial\xi^2}\overset{n}{\underset{1}{\Sigma}}\Psi}}$, and the distribution of ξ is normal.

For example, in the normal curve, for which

$$w = \frac{1}{c\sqrt{\pi}} e^{-\frac{(x-x_t)^2}{c^2}} = \frac{1}{\sqrt{\pi}} e^{-\log c - \frac{(x-x_t)^2}{c^2}},$$

we write $\overset{n}{\underset{1}{\Sigma}}\Psi = -n\log(c_0+\gamma) - \dfrac{\Sigma(x_0+\xi-x_t)^2}{(c_0+\gamma)^2}$.

$\dfrac{\partial}{\partial\xi}\Sigma\Psi = -\dfrac{2\{n(x_0+\xi)-n\bar{x}\}}{(c_0+\gamma)^2}$, and when $\xi = 0$, $x_0 = \bar{x}$, the average of $x_1 \ldots x_t \ldots x_n$.

$\dfrac{\partial}{\partial\gamma}\Sigma\Psi = -\dfrac{n}{c_0+\gamma} + \dfrac{2\Sigma(x_0+\xi-x_t)^2}{(c_0+\gamma)^3}$, and when $\xi = 0 = \gamma$ and $x_0 = \bar{x}$, $c_0^2 = \dfrac{2\Sigma(\bar{x}-x_t)^2}{n}$.

$\dfrac{\partial^2}{\partial\xi\,.\,\partial\gamma}\Sigma\Psi = \dfrac{4\{n(x_0+\xi)-n\bar{x}\}}{(c_0+\gamma)^3}$, and this is zero when $\xi = 0$ and $x_0 = \bar{x}$.

$\dfrac{\partial^2}{\partial\xi^2}\Sigma\Psi = \dfrac{-2n}{(c_0+\gamma)^2} = \dfrac{-2n}{c_0^2}$ when $\gamma = 0$.

$\dfrac{\partial^2}{\partial\gamma^2}\Sigma\Psi = \dfrac{n}{(c_0+\gamma)^2} - \dfrac{6\Sigma(x_0+\xi-x_t)^2}{(c_0+\gamma)^4} = \Bigl($when $\xi = 0 = \gamma$, $x_0 = \bar{x}$ and $c_0^2 = \dfrac{2\Sigma(\bar{x}-x_t)^2}{n}\Bigr)$, $\dfrac{n}{c_0^2} - \dfrac{3nc_0^2}{c_0^4} = -\dfrac{2n}{c_0^2}$.

Hence the chance that the observations will be made if the curve had centre $x_0+\xi$ and modulus $c_0+\gamma$ is $P = P_0 e^{-\frac{n\xi^2}{c_0^2}-\frac{n\gamma^2}{c_0^2}}$, as already found.

It can readily be seen that the required values of $\frac{\partial^2}{\partial\xi^2}\Sigma\Psi$, $\frac{\partial^2}{\partial\alpha^2}\Sigma\Psi$ . . . can be more rapidly obtained by putting $x = x_0$, $a = a_0 \ldots$ in $\frac{\partial^2}{\partial x^2}, \frac{\partial^2}{\partial\alpha^2} \ldots$ applied to $\Sigma\Psi(x - x_t, y - y_t, \ldots a, b \ldots)$.

Standard Deviation of One Mean: General Case

Now $\Psi \equiv \log f - \log H$, and $\frac{\partial}{\partial x}\Psi = \frac{\partial}{\partial x}\log f$.

Write σ_x for the standard deviation of the quantity, x, taken alone.

Then the chance P that $x_1 \ldots x_t \ldots x_n$ would be drawn from a frequency group $w = f(x - x_t)$ for which the x quantity is $x_0 + \xi$, is $P = P_0 e^{-\frac{\xi^2}{2\sigma_x^2}}$, when x_0 is that value of x which makes P greatest, P_0 is the resulting maximum value of P, and

$$\frac{1}{\sigma_x^2} = -\frac{\partial^2}{\partial x^2}\sum_1^n \log f(x - x_t) = -\sum_1^n \frac{\partial^2}{dx^2}\log f(x - x_t).$$

$$= -n \times \text{Mean}\ \frac{\partial^2}{\partial x^2}\log f(x - x_t).$$

If, as is generally the case, this mean value is finite, σ_x is of order $n^{-\frac{1}{2}}$.

As n is increased this mean calculated from the observations approximates (presumably to the same order of accuracy, viz. the neglect of terms in $n^{-\frac{1}{2}}$) to the corresponding mean expressed in terms of the constants of the curve.* The equation of the curve may be written $w = f(x)$, and the mean is then

$$\int_{-\infty}^{\infty} w \cdot \frac{d^2}{dx^2}(\log w) \cdot dx \div \int_{-\infty}^{\infty} w \cdot dx,$$

that is, each value of $\frac{\partial^2}{\partial x^2}\log w$ is given the frequency, w, with which it occurs in the curve, and the result is divided by the area of the curve, which may be taken as unity; *i.e.*

$$\int_{-\infty}^{\infty} w\, dx = 1.$$

* Up to this point the analysis is not given explicitly by Edgeworth. From here onwards we are definitely on his lines (see *J.S.S.*, 1908, pp. 662 *seq.*, and 1909, pp. 81 *seq.*).

Now

$$\int_{-\infty}^{\infty} w\frac{d^2}{dx^2}(\log w)dx = \left[w \cdot \frac{d}{dx}(\log w)\right]_{-\infty}^{\infty} - \int_{-\infty}^{\infty} \frac{dw}{dx} \cdot \frac{d}{dx}(\log w)dx$$
$$= \left[f'(x)\right]_{-\infty}^{\infty} - \int_{-\infty}^{\infty} \frac{1}{w}\left(\frac{dw}{dx}\right)^2 dx.$$

In the case, generally assumed, that the frequency axis of x is asymptotal to the curve, $f'(\infty) = 0 = f'(-\infty)$, and finally,

$$\sigma_x = \frac{1}{\sqrt{n}} \div \sqrt{\left(\int_{-\infty}^{\infty} \frac{1}{w}\left(\frac{dw}{dx}\right)^2 dx\right)}.$$

E.g. in the normal curve $w = \frac{1}{\sigma\sqrt{2\pi}}e^{-\frac{x^2}{2\sigma^2}}$, $\frac{dw}{dx} = -\frac{x}{\sigma^2}w$,

$$\int_{-\infty}^{\infty} \frac{1}{w}\left(\frac{dw}{dx}\right)^2 dx = \int_{-\infty}^{\infty} \frac{x^2w}{\sigma^4}dx = \frac{1}{\sigma^2}, \text{ and } \sigma_x = \frac{\sigma}{\sqrt{n}}.$$

Standard Deviation of the Modulus

The analysis is similar if, instead of with x, a mean of the variables, we deal singly with a secondary constant. Edgeworth takes the case of a parameter, c, a quantity which gives the scale to x and occurs only in the form $\frac{x}{c}$. The frequency curve may then be written $w\,dx = H\frac{dx}{c}e^{\Psi\left(\frac{x}{c}\right)}$.

Then P, the chance of the n selections $x_1 \ldots x_t \ldots x_n$, $= H^n c^{-n} e^{\Sigma\Psi\left(\frac{x_t}{c}\right)}$. c_0, the value of c that maximises probability, is given by $\frac{n}{c_0} + \frac{1}{c_0^2}\Sigma x_t \,.\, \Psi'\left(\frac{x_t}{c_0}\right) = 0$, and if P_0 is the corresponding maximum of P, and $c = c_0 + \gamma$,

$$\log P = \log P_0 + \tfrac{1}{2}\gamma^2 \cdot \frac{\partial}{\partial c^2}\left\{-n\log c + \Sigma\Psi\left(\frac{x_t}{c}\right)\right\} \text{ when } c_0$$

is written for c after differentiation,

$$= \log P_0 + \tfrac{1}{2}\gamma^2\left\{\frac{n}{c_0{}^2} + \frac{2}{c_0{}^3}\Sigma x_t\Psi'\left(\frac{x_t}{c}\right) + \frac{1}{c_0{}^4}\Sigma x_t{}^2\Psi''\left(\frac{x_t}{c}\right)\right\}.$$

Eliminate $\Sigma x_t\Psi'\left(\frac{x_t}{c}\right)$, and we obtain

$$\log P = \log P_0 + \tfrac{1}{2}\gamma^2\left\{-\frac{n}{c_0{}^2} + \frac{1}{c_0{}^4}\Sigma x_t{}^2\Psi''\left(\frac{x_t}{c}\right)\right\}$$

$$P = P_0 e^{-\frac{n}{2}\cdot\frac{\gamma^2}{c_0{}^2}\left\{1 - \frac{1}{c_0{}^2}\text{ Mean } x_t{}^2\Psi''\left(\frac{x_t}{c_0}\right)\right\}}.$$

Replace the mean by an integral $\int_{-\infty}^{\infty} wx^2\Psi''\left(\frac{x}{c_0}\right)dx$.

Then, since $\log w = \log H - \log c + \Psi\left(\frac{x}{c}\right)$,

$$\frac{d}{dx}(\log w) = \frac{1}{c}\Psi'\left(\frac{x}{c}\right), \text{ and } \frac{d^2}{dx^2}(\log w) = \frac{1}{c_0^2}\Psi''\left(\frac{x}{c_0}\right),$$

when c is constant and given the value c_0.

Write σ_c for the standard deviation of γ,

$$\frac{1}{\sigma_c^2} = \frac{n}{c_0^2}\left\{1 - \int_{-\infty}^{\infty} wx^2\frac{d^2}{dx^2}(\log w)dx\right\}.$$

But

$$\int_{-\infty}^{\infty} wx^2\frac{d^2}{dx^2}(\log w)dx = \left[wx^2\frac{d}{dx}(\log w)\right]_{-\infty}^{\infty}$$

$$- \int_{-\infty}^{\infty} x^2\frac{dw}{dx}\cdot\frac{d}{dx}(\log w)dx - \int_{-\infty}^{\infty} 2xw\frac{d}{dx}(\log w)dx.$$

$$= \left[x^2\frac{dw}{dx}\right]_{-\infty}^{\infty} - \int_{-\infty}^{\infty}\frac{x^2}{w}\left(\frac{dw}{dx}\right)^2dx - 2\int_{-\infty}^{\infty} x\frac{dw}{dx}\,.\,dx$$

$$= \left[x^2\frac{dw}{dx}\right]_{-\infty}^{\infty} - \int_{-\infty}^{\infty}\frac{x^2}{w}\left(\frac{dw}{dx}\right)^2dx - 2\left[xw\right]_{-\infty}^{\infty} + 2\int_{-\infty}^{\infty} w\,.\,dx.$$

Here $\int_{-\infty}^{\infty} w\,dx = 1$, and in exponential frequency curves generally $x^2\frac{dw}{dx}$ and xw are zero when $x = \pm\infty$.

Assembling the results we have

$$\frac{1}{\sigma_c^2} = \frac{n}{c_0^2}\left\{\int_{-\infty}^{\infty}\frac{1}{w}\,.\,x^2\left(\frac{dw}{dx}\right)^2dx - 1\right\}.$$

The equations now given measure the frequency of the fluctuations of the primary and secondary constants about their most probable values, when *one* set of n observations is given, and the question is from what curve may they be expected to have been drawn (*observations déjà faites*). In Edgeworth's metaphor ((51), p. 668) we are given the shot-marks on a target at which a gun was aimed, and asked what was the position of the gun. The same expressions for the fluctuations are obtained if we put a converse problem: Given the position of the gun, what will be the dispersion (*observations non faites encore*) of the average (or some other assigned function) of a *number of sets* each of n shots? This can be seen sufficiently if we take the case of only one primary constant. The solution just given (p. 19) is that if $x = x_0$ is the position from which the observations would come with least improbability, P_0 the chance in this case, and P the chance that they would come when $x = x_0 + \xi$, then

$$\log \frac{P}{P_0} = -\frac{\xi^2}{2\sigma_x^2} = \xi^2 \times \frac{n}{2}\int_{-\infty}^{\infty} w \frac{d}{dx^2} (\log w) \,.\, dx$$

when n is great.

Hence, since the equation is symmetrical in ξ, the chance that the set for which the most probable value of x is $x_0 + \xi$ would be drawn from the set where $x = x_0$, equals the chance that the set for which the most probable value of x is x_0 would be drawn from the set where $x = x_0 + \xi$. Now in the alternative method (given the position of the gun), we suppose that the most probable value of x is determined afresh from each set of marks. It follows that in the long run the frequency curve of ξ is identical for the two methods and its standard deviation the same, always supposing n to be so large that we can neglect factors involving $n^{-\frac{1}{2}}$.

Edgeworth, in fact, wavers between these two problems ((51), pp. 385 and 662).

The Greater Precision of the Genuine Inverse Method

The solution thus given for the " genuine inverse method " is contrasted with the determination of the fluctuation by more summary methods, when the whole of the original data is not used, or where the objective is taken to be the arithmetic average, the median, or some other function of the observations, which is not proved to give the most probable result. It is shown *a priori* and generally ((51), pp. 506–7) that the fluctuation of the *quæsitum* determined by the genuine inverse method is less than that of any other determination (unless the two determinations yield the same formula). This general proposition is illustrated by comparing the " genuine " formula for standard deviation of a primary constant with the usual formula $\sqrt{\frac{\mu_2}{n}}$ for the standard deviation of an arithmetic mean, and of a secondary constant, the modulus, with the formula $\sqrt{\frac{(\mu_4 - \mu_2^2)}{4n}}$. The rival formulæ are those obtained by computing the fluctuation of the arithmetic average and of the standard deviation, when many sets of n are drawn from a frequency curve where average and moments (μ_2, μ_4) are given.*

For the proofs we need the following results, based on Schwartz's Theorem, which may be expressed as follows :—

If u and v are functions of x, then, as in the well-known example of the correlation coefficient,

$$\text{Mean } u^2 \times \text{Mean } v^2 > (\text{Mean } uv)^2, \quad . \quad . \quad . \quad . \quad . \quad . \quad . \quad . \quad (a)$$

unless $u = kv$, k constant, when $>$ is replaced by $=$.

Also generally Mean $u^2 - (\text{Mean } u)^2 =$

$$\text{Mean}\{u - (\text{Mean } u)\}^2 \quad . \quad . \quad . \quad . \quad (b)$$

Further, in the frequency curves $w = f(x)$ under consideration,

$$\int_{-\infty}^{\infty} w \,.\, dx = 1, \quad \int_{-\infty}^{\infty} w \,.\, F(x) \,.\, dx = \text{Mean } F(x), \text{ where } F(x) \text{ is any integrable function} \quad . \quad . \quad . \quad . \quad (c)$$

$$\text{and Mean}\left[\frac{x}{w} \cdot \frac{dw}{dx}\right] = \int_{-\infty}^{\infty} x\,\frac{dw}{dx} \,.\, dx = \Big[xw\Big]_{-\infty}^{\infty} - \int_{-\infty}^{\infty} w \,.\, dx$$
$$= -1 \quad . \quad (d)$$

$$\frac{\sigma_x^{-2}}{n}, \text{ as above determined, } = \int_{-\infty}^{\infty} \frac{1}{w}\left(\frac{dw}{dx}\right)^2 dx = \text{Mean}\left[\left(\frac{1}{w} \cdot \frac{dw}{dx}\right)^2\right]$$

by (c).

Write $\sigma_{\bar{x}}$ as for the standard deviation of an average as determined by the Pearsonian method, when the variation of the average of samples from a given universe is considered.

Then $n\sigma_{\bar{x}}^{2} = \int_{\infty}^{\infty} w\, x^2 dx = \text{Mean } x^2$, where x is measured from the average of the frequency curve.

* The comparison of the reciprocal of the fluctuation in the form $\frac{d^2}{dx^2}\sum_1^n \log f(x - x_t)$ with that obtained by operating only on the averages of observations instead of on the original data is not illustrated in detail. The whole discussion is connected with the criticism of the method of least squares, which, operating on the average instead of on the most probable value, does not (except when distribution is normal) give the most probable result or that of least detriment. (See p. 99.) But if the equation $\frac{d}{dx}\sum_1^n \log f(x - x_t) = 0$ cannot be solved, then least squares must be used *faute de mieux*.

Then by (a), Mean $x^2 \times$ Mean $\left(\frac{1}{w} \cdot \frac{dw}{dx}\right)^2 >$ Mean $\left(\frac{x}{w} \cdot \frac{dw}{dx}\right)^2$, that is > 1, by (d).

$$\therefore \quad n\sigma_{\bar{x}}^2 \times \frac{\sigma_x^{-2}}{n} > 1,$$

$$\therefore \quad \sigma_x^2 < \sigma_{\bar{x}}^2,$$

unless $\frac{1}{w} \cdot \frac{dw}{dx} = kx$ and $w = Ce^{-\frac{1}{2}kx^2}$ (where k and C are constant), that is unless the frequency curve is normal.*

If the frequency curve is symmetrical the position for maximum probability is the average of the curve, and therefore, in non-normal symmetrical curves, determination by the genuine inverse method gives a smaller fluctuation for the average than by the alternative method.

Next compare the standard deviation (σ_c) of a parameter given on p. 21, with that of the standard deviation (σ_σ) by the alternative method. (Cf. (51), *J.S.S.*, 1909, pp. 81–2.)

$$4n\sigma_\sigma^2 = \text{Mean } x^4 - (\text{Mean } x^2)^2 = \text{Mean}\{x^2 - (\text{Mean } x^2)\}^2 \text{ by } (b)$$
$$= \text{Mean } (u^2), \text{ say.}$$

$$\frac{\sigma_c^{-2}}{n} = \frac{1}{c_0^2}\left\{\text{Mean}\left(\frac{x}{w}\frac{dw}{dx}\right)^2 - 1\right\}, \text{ by p. 21 and } (c)$$

$$= \frac{1}{c_0^2}\left[\text{Mean }\left(\frac{x}{w}\frac{dw}{dx}\right)^2 - \left\{\text{Mean }\left(\frac{x}{w}\frac{dw}{dx}\right)\right\}^2\right],$$

$$\text{since Mean }\left(\frac{x}{w}\frac{dw}{dx}\right) = -1, \text{ by } (d)$$

$$= \frac{1}{c_0^2} \text{ Mean }\left\{\frac{x}{w}\frac{dw}{dx} - \text{Mean }\left(\frac{x}{w}\frac{dx}{dw}\right)\right\}^2 \text{ by } (b)$$
$$= \text{Mean } (v^2), \text{ say.}$$

Then $4n\sigma_\sigma^2 \times \frac{\sigma_c^{-2}}{n} = \text{Mean } (u^2) \times \text{Mean } (v^2) > \{\text{Mean } (uv)\}^2$ by (a)

$$\text{Mean } (uv) = \text{Mean}\left\{(x^2 - \sigma^2)\left(\frac{x}{w}\frac{dw}{dx} + 1\right)\right\} \times \frac{1}{c_0^2} \text{ by } (d)$$

$$c_0^2 \times \text{Mean } (uv) = \text{Mean } \frac{x^3}{w}\frac{dw}{dx} - \sigma^2 \text{ Mean}\left(\frac{x}{w}\frac{dw}{dx}\right) + \text{Mean } x^2 - \sigma^2.$$

* See (51), pp. 674–6 with reference to pp. 505–6 for proof that the results by the genuine inverse method and by the summary method of dealing with the Arithmetic Mean are identical for the Generalised Law of Error. The breach of the equation $\frac{1}{w} \cdot \frac{dw}{dx} = kx$ is apparently justified by the neglect of certain small quantities.

Now Mean $\left(\frac{x^3}{w}\frac{dw}{dx}\right) = \int_{-\infty}^{\infty} x^3 \frac{dw}{dx} dw$, by (c), $= \left[x^3 w\right]_{-\infty}^{\infty}$ $- 3\int_{-\infty}^{\infty} wx^2 dx = 0 - 3\sigma^2$, if $x^3 w$ is 0 when $x = \pm\infty$, as may be assumed;

$$\therefore\ c_0{}^2 \text{ Mean } (uv) = -3\sigma^2 + \sigma^2 + \sigma^2 - \sigma^2 = -2\sigma^2.$$

We have, therefore, $4\sigma_\sigma{}^2 \times \sigma_c{}^{-2} > \left(\frac{-2\sigma^2}{c_0{}^2}\right)^2$

$$\left(\frac{\sigma_c}{c_0{}^2}\right)^2 < \left(\frac{\sigma_\sigma}{\sigma^2}\right)^2$$

and the *relative* deviation of the parameter by the genuine inverse method is less than that of the standard deviation by the alternative method,

unless $u = kv$, i.e. $x^2 - \sigma^2 = \left(\frac{x}{w}\frac{dw}{dx} + 1\right)\frac{k}{c_0}$ for all values of x;

and since this includes the case $x = 0$, we have $k = -c_0\sigma^2$, and $w = Ae^{\frac{-x^2}{2\sigma^2}}$, where A is constant; that is, we have the normal law. [This condition is not given explicitly by Edgeworth.]

A further illustration is given ((51), pp. 676–8) by Pearson's Type III, which may be written

$$y = Hx^p e^{-\gamma x}.$$

The standard deviation for the mode is given by the equation $\frac{1}{n\sigma^2} = \int_{-\infty}^{\infty} \frac{1}{y}\left(\frac{dy}{dx}\right)^2 dx$, by the formula on p. 20, which when worked out becomes $\frac{\gamma^2}{(p-1)}$, since $\int_{-\infty}^{\infty} y \,.\, dx = 1$. The second moment about the average is

$$\int_{-\infty}^{\infty} yx^2 dx - \left(\int_{-\infty}^{\infty} yx dx\right)^2 = \frac{p+1}{\gamma^2}$$

and the standard deviation of the arithmetic average by the alternative method is therefore given by

$$\sigma_{\bar{x}}{}^2 = \frac{p+1}{n\gamma^2}$$

$$\therefore\ \sigma^2 = \frac{p-1}{p+1}\sigma_{\bar{x}}{}^2,$$

and $\sigma < \sigma_{\bar{x}}$, unless p is infinite and the curve is normal.

Summary

The principal results of this analysis may be thus summed up :—

Let $w = \frac{H}{c} e^{\psi\left(\frac{x}{c}\right)}$ be the form of a frequency curve, satisfying the usual conditions, with one variable and one parameter. Then the method of genuine Inverse Probability shows that the best values for the mean size of x and for c are given by x_0 and c_0, where

$$\Sigma\psi'(x_t - x_o) = 0, \text{ and } \frac{n}{c_0} + \frac{1}{c_0^2}\Sigma x_t\psi'\left(\frac{x_t}{c}\right) = 0,$$

the summation extending over n observed values of x, viz. $x_1 \ldots x_t \ldots x_n$, where n is large; and that the resulting fluctuations * of x_0 and c_0,

$$\text{viz. } 1 \div nR, \text{ and } 1 \div n(S-1),$$

$$\text{where } R = \int \frac{1}{w}\left(\frac{dw}{dx}\right)^2 dx,\ S = \int \frac{x^2}{w}\left(\frac{dw}{dx}\right)^2 dx,$$

are less than those of any other mean of the observations or value of the parameter.

Alternative Methods

Edgeworth proceeds to show ((51), *J.S.S.*, 1909, pp. 82–90) that these results can be obtained by " a direct method free from the speculative character which attaches to inverse probability." Instead of mounting from *observations déjà faites* to the source, let us watch the fluctuations resulting from a series of n-fold observations (*observations non faites encore*) made at random from a frequency curve with constants assumed to be known.

With the equation of the frequency curve as above, let $\theta(x_t - \chi)$ be an unknown function of a deviation of an observation from χ, where χ belongs to the comprehensive class of functions (of the quantities $x_1 \ldots x_t \ldots$) that constitute averages, defined by the equation $\Sigma\theta(x_t - \chi) = 0$; for example, χ might be given by the equation $\Sigma(x_t - \chi)^p = 0$.

It is required to find the form of θ, which yields the value of χ that has least fluctuation.

It may be assumed that, n being large, the " fluctuation "

* This term is used here for the square of the standard deviation.

of χ is of the order n^{-1}; so that, if χ_0 is a mean value, and $\xi = \chi - \chi_0$, terms in ξ^2 are of order n^{-1} and may be neglected in comparison with terms in ξ.

Take the parameter c as unity, a simplification that does not affect the result.

Write $u_t = x_t - \chi_0$ and so replace $x_t - \chi$ by $u_t - \xi$.

Then $0 = \Sigma\theta(u_t - \xi) = \Sigma\theta(u_t) - \xi\Sigma\theta'(u_t) +$ negligible terms.

It is known, as an example of a general formula, that the fluctuation of $\Sigma\theta(u_t)$ is $\Sigma\{\theta(u_t)\}^2 - (\Sigma\theta(u_t))^2$, and in this case the second term tends to zero.

Replace summation by integration; thus $\Sigma\{\theta(u_t)\}^2 =$ approximately Mean $\{\theta(u)\}^2 wdu \times n = \int\{\theta(u)\}^2 wdu \times n$, since $\int wdu$ may be taken as unity. Similarly $\Sigma\theta'(u_t) = \int\theta'(u)wdu \times n$.

Then the fluctuation of ξ equals the fluctuation of $\Sigma\theta(u_t)$ divided by $[\Sigma\theta'(u_t)]^2$, whose variation is negligible,

$$= \frac{n\int\{\theta(u)\}^2 wdu}{n^2[\int\theta'(u)wdu]^2} = \frac{P^2}{nQ^2},$$

where P^2 and Q^2 are definite integrals equal respectively to the numerator and denominator divided by n and n^2.

The form of θ which minimises this expression is obtained by writing $\theta + \delta\theta$ for θ, expanding in powers of $\delta\theta$, and equating the coefficient of $\delta\theta(u)$ to 0.

Observing that

$$\int_{-\infty}^{\infty} \delta\{\theta'(u)\}wdu = \Big[w\delta\{\theta(u)\}\Big]_{-\infty}^{\infty} - \int_{-\infty}^{\infty}\frac{dw}{du}\delta\{\theta(u)\}du$$
$$= - \int_{-\infty}^{\infty}\frac{dw}{du}\,\delta\{\theta(u)\}du,$$

we have, after expanding and collecting terms

$$Q\int w\theta(u)du + P^2\int\frac{dw}{du}du = 0.$$

This equation is satisfied by $\theta \equiv -\dfrac{P^2}{Q}\cdot\dfrac{1}{w}\cdot\dfrac{dw}{du}$, where $-\dfrac{P^2}{Q}$ is constant, say A.

Then $P^2 = \displaystyle\int A^2\frac{1}{w^2}\Big(\frac{dw}{du}\Big)^2 wdu = A^2\int\frac{1}{w}\Big(\frac{dw}{dx}\Big)^2 dx = A^2R$, where R is the expression on p. 26, and $Q =$

$$\int\frac{d}{du}\Big(\frac{A}{w}\cdot\frac{dw}{du}\Big)wdu = \Big[\frac{A}{w}\cdot\frac{dw}{du}w\Big]_{-\infty}^{\infty} - A\int_{-\infty}^{\infty}\frac{1}{w}\Big(\frac{dw}{du}\Big)^2 wdu = -AR.$$

It is shown that this results in a minimum, not a maximum,

for the fluctuation of ξ, which is identical with the fluctuation of χ.

Hence the form for θ which makes the fluctuation of χ a minimum is $\theta = A \cdot \frac{1}{w}\frac{dw}{dx} = A\frac{d \log w}{dx} = A\psi'(x)$, since if $c = 1$ $w = He^{\psi(x)}$.

Hence χ is given by $0 = A\Sigma\psi'(x_t - \chi)$,

$$\therefore \quad 0 = \Sigma\psi'(x_t - \chi),$$

that is by the equation which gives x_0 by the method of Inverse Probability.

The fluctuation of ξ is $\frac{P^2}{nQ^2} = \frac{A^2R}{nA^2R^2} = \frac{1}{nR}$, which is also the same as obtained before.

By a more intricate analysis it is shown that the same values for the best parameter and its fluctuation are obtained by the two methods.

Standard Deviation of Average when the Frequency Form is Unknown

These determinations depend, of course, on our knowledge of the form of the frequency curve. If we are solely dependent on the observations we must follow a less exact method ((51), p. 510–11).

Suppose the n observations broken up into γ groups each containing m observations, so that $\gamma m = n$.

Write $\xi_1 = (x_1 + x_2 + \ldots + x_m)/m$,

$$\xi_2 = (x_{m+1} + x_{m+2} + \ldots x_{2m})/m, \text{ and so on.}$$

Then if m is great the ξ's tend to be distributed normally, with average equal to the average of the x's and fluctuation σ_ξ^2, given by $m\sigma_\xi^2 = \sigma_x^2$.

Then the fluctuation of the average of the ξ's is given by

$$\sigma_1^2 = \frac{1}{\gamma}\sigma_\xi^2 = \frac{1}{\gamma m}\sigma_x^2 = \frac{1}{n}\sigma_x^2$$

and therefore equals the fluctuation of the average of the x's.

Hence, if we work with the original observations, non-normal or not known to be normal, we can still say, if n is great, that the fluctuation of the average is approximately $\frac{\sigma_x^2}{n}$, and that the normal table of chances is applicable (see next chapter).

CHAPTER III

THE LAW OF ERROR

THE law of error is the subject of the first of the papers listed in the bibliography and also of the last, and there is hardly a topic or application of statistics in the intermediate papers which does not depend on it directly or is not related to the conception of probability on which it is based.

Two or more lines of inquiry are followed. The more distinctive to the author are " What, if any, is the law of probability, of the frequency of different values for a mean ? " and " How general is the conception of a mean to which this law applies ? " These inquiries are purely mathematical and subjective, and do not depend on any realisation in the objective universe. A separate investigation, continually and naturally entangled in the analysis, relates to the representation of statistics (arising from games of chance, errors in observation, physical, biological, and economic distributions) by mathematical formulæ, which either express the generalised law of error itself or some allied form, and the relationship of such formulæ to Prof. Karl Pearson's system is frequently discussed. The subject of correlation takes its place in the papers from 1892 onwards.

The opening sentence of the study which virtually completes the main line of investigation is, " The law which is the subject of this paper is the approximate expression for the frequency with which different values are assumed by a magnitude that depends on a number of independently varying magnitudes, here called elements " ((47), p. 36).

In the earlier classical writings on the subject these elements were the errors (instrumental defects, atmospheric disturbance, etc.) whose sum resulted in the magnitude, " error of observation." Each elemental error was conceived as taking any small value by chance, the probability of each value being given by the ordinate of a facility curve. This statement is extended in various ways in the sentence just quoted. The elements are not commonly errors, but any quantities, such as the number of successes in an ideal game of chance, small disturbances or deviations whose aggregate makes a measured or measurable quantity, the deviations from the centre in any frequency group, etc., where the total or the average determined by extraction of samples is the major magnitude. The dependence of the magnitude on the elements is not necessarily that of simple summation or averaging, but may be a weighted sum or any function, subject to limitations discussed in the analysis. The elemental curves and the resulting frequency of the magnitudes may be replaced by elemental and resulting surfaces in two or more dimensions. It should be added that fortuity is not rigidly necessary either in the supposed selection from any one elemental group, or in the aggregation of the elements to form the major magnitude.

The nature of the approximation depends on the convergency of a series expanded in terms of $m^{-\frac{1}{2}}$, where m is the number of elements the sum or other function of which forms the magnitude. The first term, the limit when m is infinite, is the Gaussian law of error. If the second term is included (terms which contain $\frac{1}{m}$ with finite coefficients being neglected) the approximation is the unsymmetric law of error as given by Laplace and Poisson. The inclusion of the third term (which is symmetrical) introduces the measurement of kurtosis (in Prof. Pearson's nomenclature), but the approximation is not identical with Prof. Pearson's system.

The coefficients of powers of m in the successive terms depend on the shape of the elemental facility curves, more exactly on the divergence of the moments of these curves from what they would be if the curves were normal. Thus j, the constant coefficient involving $m^{-\frac{1}{2}}$, in the approximation, is akin to the mean of the measurements of asymmetry (Prof. Pearson's $\sqrt{\beta_1}$) in the elemental curves multiplied by $m^{-\frac{1}{2}}$, and i, the constant coefficient involving m^{-1} in the third approximation is akin to the mean of the excess of their measurements of kurtosis over 3, *i.e.* $\beta_2 - 3$ in the same notation, multiplied by m^{-1}.

The elemental curves are limited by the condition that such measurements of abnormality (the measurements already named and others depending upon higher moments) should all be finite, such that, when multiplied by the appropriate power of $m^{-\frac{1}{2}}$ and by other coefficients involving the deviation (x) of the major magnitude from its average, the products should be negligible.

It is found that as we proceed along the abscissa of the resulting curve of error, the terms neglected in successive approximations become more important and the series tends to lose its convergence. Closeness of approximation is to be assumed only in the central part of the curve, this "central" part including, for values of the constants assumed to be typical, about ·85 of the area of the curve. Beyond some such position the approximation is less and less exact, till no regular shape appears at all.

The Equation of the Law of Error

The results may thus be summarised; the proofs follow later:

The equation to the normal curve is in its general form

$$z = e^{-\frac{1}{3!}k_1D^3 + \frac{1}{4!}k_2D^4 - \cdots}\left(\frac{1}{c\sqrt{\pi}}e^{-\frac{x^2}{c^2}}\right)$$

where c is the modulus of the curve, and k_1, k_2 . . . are constants depending on the successive moments of the curve and are of descending magnitude in powers of an unknown $m^{-\frac{1}{2}}$ (unknown unless the actual genesis of the curve is known), and D represents differentiation with regard to x.

The successive approximations are (1) $\frac{1}{c\sqrt{\pi}}e^{-\frac{x^2}{c^2}}$;

(2) the expression obtained by using k_1 and neglecting k_2, etc.;

(3) including k_2 also; and so on. The third approximation worked out is

$$z = \frac{1}{c\sqrt{\pi}}e^{-\frac{x^2}{c^2}}\left\{1 - j\left(2\frac{x}{c} - \frac{4}{3}\cdot\frac{x^3}{c^3}\right) + i\left(\frac{1}{2} - 2\frac{x^2}{c^2} + \frac{2}{3}\cdot\frac{x^4}{c^4}\right)\right.$$
$$\left. + j^2\left(-\frac{5}{3} + 10\frac{x^2}{c^2} - \frac{20}{3}\cdot\frac{x^4}{c^4} + \frac{8}{9}\cdot\frac{x^6}{c^6}\right)\right\},$$

where $c^2 = 2\int_{-\infty}^{\infty} x^2 z dx$, $jc^3 = \int_{-\infty}^{\infty} x^3 z dx$, and $ic^4 = \int_{-\infty}^{\infty} x^4 z dx - \frac{3}{4}$.

The origin of the abscissæ is that of the centre of gravity of the curve. The second approximation is obtained by omitting the i, j^2 terms.

In following the mathematics and studying the formulæ it is

necessary for modern readers to become accustomed to the use of the modulus, c, or its reciprocal, h, instead of the standard deviation ($c = \sigma\sqrt{2}$). The latter was not in general use at the date of the earlier papers, and the notation of least squares used the symbol here called h. The forms of the functions and the work of reduction are in fact very often simpler when the modulus is used and τ written for $\frac{x}{c}$. It has seemed better not to re-write the equations in general in terms of σ; there is no difficulty in transforming them if tables are used in which σ is the unit.

The integral form of the equation just written is

$$\int_0^x z dx = \int_0^x \frac{1}{c\sqrt{\pi}} e^{-\frac{x^2}{c^2}} dx - \frac{j}{3\sqrt{\pi}} \left\{1 - \left(1 - 2\frac{x^2}{c^2}\right) e^{-\frac{x^2}{c^2}}\right\}$$

$$+ i\left(\frac{x}{2c} - \frac{x^3}{3c^3}\right)\frac{e^{-\frac{x^2}{c^2}}}{\sqrt{\pi}} + j^2\left(-\frac{5x}{c} + \frac{20x^3}{c^3} - \frac{4x^5}{9c^5}\right)\frac{e^{-\frac{x^2}{c^2}}}{\sqrt{\pi}}.$$

The area is unity.

Write $x = \tau c$, and write z' for the resulting expression for z. Then

$$\int_0^\tau z' d\tau = \int_0^\tau \frac{1}{\sqrt{\pi}} e^{-\tau^2} d\tau - \frac{j}{3\sqrt{\pi}} \left\{1 - (1 - 2\tau^2) e^{-\tau^2}\right\}$$

$$+ i(\tfrac{1}{2}\tau - \tfrac{1}{3}\tau^3)\frac{e^{-\tau^2}}{\sqrt{\pi}} + j^2(-5\tau + 20\tau^3 - \tfrac{4}{9}\tau^5)\frac{e^{-\tau^2}}{\sqrt{\pi}}.$$

In terms of the standard deviation σ, with $\beta_1 = \frac{\mu_3^2}{\mu_2^3}$ and $\beta_2 = \frac{\mu_4}{\mu_2^2}$, the integral from 0 to x is

$$\int_0^x \frac{1}{\sigma\sqrt{2\pi}} e^{-\frac{x^2}{2\sigma^2}} dx - \frac{\sqrt{\beta_1}}{6\sqrt{2\pi}} \left[1 - \left(1 - \frac{x^2}{\sigma^2}\right) e^{-\frac{x^2}{2\sigma^2}}\right] +$$

$$(\beta_2 - 3)\left(\frac{x}{8\sigma} - \frac{x^3}{24\sigma^3}\right)\frac{e^{-\frac{x^2}{2\sigma^2}}}{\sqrt{2\pi}} + \frac{\beta_1}{8}\left(-\frac{5x}{\sigma} + \frac{10x^3}{\sigma^3} - \frac{x^5}{9\sigma^5}\right)\frac{e^{-\frac{x^2}{2\sigma^2}}}{\sqrt{2\pi}}.$$

Here $\beta_1 = 8j^2$, $\beta_2 = 3 + 4i$.

The mode of this curve is at $-jc$ and the median at $-\frac{1}{3}jc$, terms of lower order than i or j^2 being neglected. To the same order of approximation the area standing on the abscissa from the median to the mean is $\frac{j}{3\sqrt{\pi}}$, and if j is positive the area to the left of the centre of gravity exceeds that on the right by

$\frac{2}{3}j\sqrt{\pi}$. The whole chance of an error in defect of the average is greater than that of an error in excess when j is positive. $\int_{-x}^{x} z dx$ does not involve the j term and $= \int_{-x}^{x} \frac{1}{c\sqrt{\pi}} e^{-\frac{x^2}{c^2}} dx$ correct to the second approximation.

The curve is reproductive in the sense that if a variable X equals the sum of two or more variables x_1, x_2 . . . which fluctuate according to the generalised equation with moduli c_1, c_2 . . ., then X fluctuates also according to the generalised equation with modulus C, where $C^2 = c_1{}^2 + c_2{}^2 + \ldots$, and the other constants of the X curve can be expressed in terms of those of the x_1, x_2 . . . curves.

Relation to other Formulæ

The law of error given by these formulæ includes the formulæ of Bernoulli, Gauss, Laplace and Poisson and carries on the analysis due to Glaisher. Bernoulli's theorems with Poisson's extension are obtained by considering each elemental curve as consisting of two alternatives (success and failure) the chances of which are p and q $(p + q = 1)$. In this case the successive moments of each elemental curve are p, $\sqrt{pq}$, $pq(q - p)$, etc. If m independent selections are made, $c = \sqrt{2pqm}$, $j = \frac{q - p}{\sqrt{8pqm}}$, $i = \frac{1}{4} \cdot \frac{1 - 6pq}{pqm}$.* If p varies from one elemental curve to another, $c^2 = 2\Sigma pq$, $j = \frac{\Sigma pq(q - p)}{c^3}$, and $i = \frac{\Sigma pq(1 - 3pq) - 3\Sigma p^2q^2}{4(\Sigma p^2q^2)}$.

The Gaussian normal law is obtained by taking m so great that the terms involving j, etc., are negligible.

The case where m balls are taken from an urn in a group, instead of one by one with continual replacement, expressed by the hypergeometric series, is included when m is moderately large, since the amount of interdependence involved is within the permissible limits. If m is large, then in Prof. Pearson's equation, as originally stated, $\frac{1}{y} \cdot \frac{dy}{dx} = \frac{x + a}{b_0 + b_1 x + b_2 x^2}$, b_2 is of

* If p is of the order $\frac{1}{m}$, so that pm is a small finite number, the approximation is no longer convergent, and the analysis leading to Poisson's series must be substituted.

order $\frac{1}{m}$, b_1 is of order $\frac{1}{\sqrt{m}}$, and the Edgeworthian statement is obtained correct to terms in $\frac{1}{\sqrt{m}}$.

Scope of Law

The Law has a very special importance in connection with the precision and frequency curve of averages. In this case each elemental curve is the frequency curve from which m examples are drawn at random. Let $\bar{x}$, σ, μ_2, μ_3, . . . be the mean, standard deviation and moments of this curve. The constants of the frequency curve of the average are $c = \frac{\sigma\sqrt{2}}{\sqrt{m}}$, $j = \frac{1}{2\sqrt{2m}} \cdot \frac{\mu_3}{\sigma^3}$, $i = \frac{1}{4m}\left(\frac{\mu_4}{\sigma^4} - 3\right)$, and the further coefficients depend on such quantities as $\frac{1}{m^{p-1}}\left(\frac{\mu_{2p}}{\sigma^{2p}} - \frac{(2p)!}{2^p \, . \, p!}\right)$. Unless the elemental curve is extremely abnormal the Edgeworthian series is rapidly convergent.

If the drawing to be averaged is made without replacement from a group containing ν items, c becomes $\frac{c}{\sqrt{m\left(1 - \frac{m}{\nu}\right)}}$, and j, i and subsequent coefficients are modified by terms of the order $\left(\frac{m}{\nu}\right)^{\frac{1}{2}}$. This amount of interdependence, when $\frac{m}{\nu}$ is small, is shown to be permissible without further modification of the terms.

The phrase " the law of great numbers " may be taken as meaning the law of the variation of totals or averages (such as death rates) which depend on a multitude of independent items. The law indicates the approximate constancy of such totals, the coefficient of variation being small, since it involves the factor $\frac{1}{\sqrt{m}}$, and the normality of the distribution of the variations. The Edgeworthian series is a complete statement of this law, and its postulates show the conditions under which it is obtained.

The claim is definitely made that the formula

$$z = e^{-\frac{1}{3!}k_1D_3 + \frac{1}{4!}k_2D_4 + \ . \ . \ .}\left(\frac{1}{c\sqrt{\pi}}e^{-\frac{x^2}{c^2}}\right)$$

is that of the true and unique law that represents the frequency curve of a magnitude that depends on a number of independent elements, and that no other can be deduced from general *a priori* conditions. There is no claim that in general it adequately represents the phenomena which nature exhibits, for it is not known whether natural magnitudes are due to the aggregation of numerous independent elements of the kind supposed. This question can only be answered by a comparison of natural frequency curves with the law of error. But Edgeworth clearly held that any other formula must rest on an empirical or *a posteriori* basis, for the only general *a priori* basis leads to his equation.*

Whether this distribution is or is not produced in natural phenomena, Edgeworth emphasises again and again that it is applicable to averages and other functions drawn from natural frequency groups, and is therefore of fundamental importance in practical statistics, in such problems as relate to the precision of an average or the significance of a difference.

Two or more Dimensions

The extension of the theory to the correlation surface is obtained by similar analysis. The postulates are extended by replacing elemental curves by elemental surfaces, for each of which the frequency of a pair of values (ξ, η) is a function of ξ and η. The major magnitude whose frequency is sought is now the pair x, y, where x is related to the ξ's as in the former case, and y is related by an equation of the same form to the η's. The resulting law is (so far as the second approximation)

$$z = e^{-\frac{1}{3!}\left(m_{30}\frac{\partial^3}{\partial x^3} + 3m_{31}\frac{\partial^3}{\partial x^2 \partial y} + 3m_{12}\frac{\partial^3}{\partial x \partial y^2} + m_{03}\frac{\partial^3}{\partial y^3}\right)} \times \frac{1}{\pi c_1 c_2 \sqrt{1-r^2}} e^{-\frac{1}{1-r^2}\left(\frac{x^2}{c_1^2} + \frac{y^2}{c_2^2} - 2r\frac{xy}{c_1 c_2}\right)}.$$

* This claim for the uniqueness of the Edgeworthian law seems not to have been understood. Thus Mr. Elderton (*Frequency Curves and Correlation:* 1927 edition, p. 128) says, " Edgeworth used a modified form of Type A," *i.e.* of the Bruns-Charlier series, without mentioning the essential distinction here made. (See also p. 136 below.) It is his Method of Translation (pp. 65 *seq.* below) that is properly to be compared with the other equations Mr. Elderton quotes, not his generalised Law of Error.

Here

$$c_1{}^2 = \int_{-\infty}^{\infty} 2x^2 z dx,\ c_2{}^2 = \int_{-\infty}^{\infty} 2y^2 z dy,\ rc_1c_2 = \int_{-\infty}^{\infty}\int_{-\infty}^{\infty} 2xyz dx dy,$$

$$m_{30} = \int_{-\infty}^{\infty} x^3 z dx,\quad m_{21} = \int_{-\infty}^{\infty}\int_{-\infty}^{\infty} x^2 yz dx dy,$$

$$m_{12} = \int_{-\infty}^{\infty}\int_{-\infty}^{\infty} xy^2 z dx dy,\quad m_{03} = \int_{-\infty}^{\infty} y^3 z dy.$$

That is, c_1, c_2, are the moduli of the distributions in x and y, and the m's are moments of the third order.

$$m_{30} = j_1 c_1{}^3,\quad m_{03} = j_2 c_2{}^3 \text{ in the former notation.}$$

Also in terms of the moments of the elemental surfaces, which we may write μ_{20}, μ_{21}, etc.,

$$c_1{}^2 = \Sigma 2\mu_{20},\ c_2{}^2 = \Sigma 2\mu_{02},\ m_{30} = \Sigma\mu_{30},\ m_{21} = \Sigma\mu_{21},$$
$$m_{12} = \Sigma\mu_{12},\ m_{03} = \Sigma\mu_{03},\ rc_1c_2 = 2\Sigma\mu_{11}.$$

The first approximation is the well-known normal surface.

It is indicated ((47), p. 120) that a similar equation can be obtained for three or more variables, and that further approximations can be written down by analogy with those for one dimension, but Edgeworth never developed these topics.

The equation, when written out, becomes $z = z_0(1 - w)$, when $z_0 = \dfrac{1}{\pi c_1 c_2 \rho} e^{-\frac{1}{\rho^2}\left(\frac{x^2}{c_1{}^2} + \frac{y^2}{c_2{}^2} - 2r\frac{xy}{c_1c_2}\right)}$, ρ^2 being written for $1 - r^2$, and w (the correcting factor to give the second approximation)

$$= \frac{2}{\rho^3}\left\{\frac{m_{30}}{3c_1{}^3}(3q - 2q^3) + \frac{m_{21}}{c_1{}^2c_2}(q' - 2rq - 2q'q^2)\right.$$
$$\left. + \frac{m_{12}}{c_1c_2{}^2}(q - 2rq' - 2qq'^2) + \frac{m_{03}}{3c_2{}^3}(3q' - 2q'^3)\right\},$$

where q, q', are written for $\left(\dfrac{x}{c_1} - r\dfrac{y}{c_2}\right)\Big/\rho$ and $\left(\dfrac{y}{c_2} - r\dfrac{x}{c_1}\right)\Big/\rho$.

The most convenient form for integration is obtained by transferring to principal axes with unit moduli. Write

$$X = \left(\frac{x}{c_1} + \frac{y}{c_2}\right)\Big/\sqrt{2(1+r)},\quad Y = \left(-\frac{x}{c_1} + \frac{y}{c_2}\right)\Big/\sqrt{2(1-r)},$$

and I_{XY} for the surface standing on the rectangle bounded by the axes of co-ordinates and parallels making intercepts X, Y.

$$\text{Then}\quad I_{XY} = \frac{1}{\rho^2\pi}\int_0^X e^{-X^2}dX \,.\int_0^Y e^{-Y^2}dY$$

$$-\frac{1}{3\rho^2\pi}\Big[M_{30}\int_0^Y e^{-Y^2}dY\,.\{1-(1-2X^2)e^{-X^2}\}$$

$$+3M_{21}e^{-X^2}\,.\,X(1-e^{-Y^2})+3M_{12}e^{-Y^2}\,.\,Y(1-e^{-X^2})$$

$$+M_{03}\int_0^X e^{-X^2}dX\,.\{1-(1-2Y^2)e^{-Y^2}\}\Big]$$

where M_{30} = Mean X^3, M_{21} = Mean X^2Y, etc.

The integral for the parallelogram bounded by $X = X_1$, $X = X_2$, $Y = Y_1$, $Y = Y_2$ is then $I_{X_2Y_2} - I_{X_1Y_2} - I_{X_2Y_1} + I_{X_1Y_1}$.

The curve of regression, taken as the locus of the *modes* of arrays in x, is

$$\frac{y}{c_2} = r\,.\,\frac{x}{c_1} + \frac{1}{\rho^2}\left\{\frac{m_{30}r}{c_1^3} - (1+2r^2)\frac{m_{21}}{c_1^2c_2} + 3r\frac{m_{12}}{c_1c_2^2} - \frac{m_{03}}{c_2^3}\right\}$$

$$-\frac{2x^2}{c_1^2}\left(r\,.\,\frac{m_{30}}{c_1^3} - \frac{m_{21}}{c_1^2c_2}\right),$$

that is, a parabola passing near, but not through, the origin.

The locus of the *means* of the arrays is the parabola

$$\frac{y}{c_2} = r\,.\,\frac{x}{c_1} + \left(1 - 2\frac{x^2}{c_1^2}\right)\left(r\,.\,\frac{m_{30}}{c_1^3} - \frac{m_{21}}{c_1^2c_2}\right).$$

It is to be noticed that $r = 2\left|\frac{x}{c_1}\,.\,\frac{y}{c_2}\right| = \left|\frac{x}{\sigma_1}\,.\,\frac{y}{\sigma_2}\right|$ is of the nature of an average,* and so comes under the ægis of the law

* (51) *Stat. Jrnl.*, 1908, pp. 510–11. Non-normal observations $x_1y_1 \ldots x_ny$ can be broken into ν groups each containing m observations. $n = \nu m$. Write $\xi_1 = (x_1 + x_2 + \ldots + x_m)/\mu$, $\xi_2 = (x_{m+1} + x_{m+2} + \ldots)/\mu \ldots$, and similarly for $\eta_1, \eta_2 \ldots\ldots$; μ is to be assigned. (See p. 28 above.) Suppose $\xi_1\eta_1, \xi_2\eta_2 \ldots$ make an approximately normal surface. Then if the double observations are independent,

$$\xi_1\eta_1 = x_1y_1 + \ldots + x_my_m. \qquad \sum_1^\nu \xi_\iota\eta_\iota = \Sigma xy.$$

Write $r_1 = \frac{\Sigma\xi_\iota\eta_\iota}{\nu\sigma_1\sigma_2}$, where σ_1 (st. dev. of ξ) $= \frac{\sqrt{m}}{\mu}\sigma_x$, $\sigma_2 = \frac{\sqrt{m}}{\mu}\sigma_y$.

Then $r_1 = \frac{\mu^2}{m\nu}\,.\,\frac{\Sigma xy}{\sigma_x\sigma_y} = \mu^2 r$, where r is the product-sum coefficient from the original observations.

Now take $\mu^2 = 1$.†

The new system in r_1 is then normal with the fluctuation found for r.

† Edgeworth omits ν in the denominator of r_1, and takes $\mu = \sqrt{m}$, so that $\sigma_1 = \sigma_x$. Apparently he also omits n in the denominator of r. He thus arrives at $r_1 = r$, as he intended.

of error. Hence, even if the elemental surfaces are so abnormal that the equation just given is not a good approximation, still the frequency curve of r as derived from sampling tends to be normal, and its standard deviation to be subject to the normal table of chance.

CHAPTER IV

PROOF OF THE LAW OF ERROR

LET there be m elemental frequency groups, such that the chance of drawing a magnitude from any, the qth, group is $\phi(a_q + \xi_q)$, where ϕ is an unknown function and a_q the average of the group.

An aggregate $A + {}_px$ is formed from magnitudes drawn independently from the m groups, say ${}_p\xi_1, {}_p\xi_2 \ldots {}_p\xi_q \ldots {}_p\xi_m$, so that quantities H_p and A are defined by

$$H_p = A + {}_px = \sum_{q=1}^{q=m} (a_q + {}_p\xi_q), \text{ and } {}_px = \sum_{q=1}^{q=m} ({}_p\xi_q).$$

It is postulated that (1) the selection of ${}_p\xi_q$ for an aggregate H_p is independent of the selection of any ${}_{p'}\xi_q$ for *another* aggregate $H_{p'}$. But the effect of partial dependence is only to increase the number of trials necessary to obtain experimentally a given closeness of approximation between observations and the law ((47), pp. 39 and 127).

(2) The drawing of ${}_p\xi_q$ is independent of the drawing of ${}_p\xi_{q'}$, the contribution of any other elemental group to the *same* aggregate. But the effect of partial dependence is only to increase the size of m, the number of elements in an aggregate, necessary for a given rapidity of convergence in the approximation ((47), pp. 39 and 126).

(3) The function ϕ must be such that $\frac{\mu_t}{\sigma^t}$ (where σ is the standard deviation and μ_t the tth moment about the average of ϕ) must differ from $\frac{(2t)!}{2^t \, . \, t!}$ (the value it would have if ϕ were normal) by a small finite number,* the admissible difference

* I am not certain whether Edgeworth went explicitly beyond the condition that $\frac{\mu_t}{\sigma_t}$ should be small and finite, but he must have been aware that it was the differences here named that entered into the equation, as he uses the fact implicitly in various places.

being the greater, the greater is m. Any elemental curves with effective range much less than that of the majority should be excluded in considering the necessary size of m. These relations become definite in the course of the analysis ((47), pp. 39 and 113–16).

The analysis relates to the simple aggregate of the elements. It can be at once extended to the weighted aggregate, say, $\Sigma w_q(a_q + {}_p\xi_q)$, by writing $w_q \,.\, {}_p\xi_q$ for ${}_p\xi_q$ in the postulates. The frequency of the average $\frac{1}{m} \,.\, H_p$, or $\frac{1}{\Sigma w} \,.\, H_p$, differs only in scale of abscissa from the frequency of the aggregate. It can also be extended, with certain limitations ((47), pp. 120 *seq.*) to the case where summation is replaced by a function which when expanded by Taylor's Theorem is rapidly convergent, to the case where the magnitude is the sum of functions of the elements, and to more complicated cases * ((48), pp. 499 *seq.*). These developments lead us to expect the appearance of the law of error in a wider field.

Following Edgeworth's "proposed new method" ((47), pp. 40–45) we proceed to determine the moments of the frequency distribution of the values of x, where any the pth value is ${}_px = \sum_{q=1}^{q=m} {}_p\xi_q$. (Evidently $A = \Sigma a_q$.)

Take any small constant quantity α, whose purpose is to enable terms of like order to be assembled and equated.

$$e^{\alpha_p x} = e^{\Sigma \alpha \,.\, {}_p\xi_q} = \prod_{q=1}^{q=m} \,.\, e^{\alpha \,.\, {}_p\xi_q},$$

where Π is the symbol for multiplying together the factors named after it.

Let $[u]$, where u is any function, be written for the mean value of u in its frequency curve.

Expand both sides of the equation and take the mean values, when p is given all possible values, using the theorems, that the mean of a sum equals the sum of the means of its items, and that the mean of a product of

* A further extension is named by some possible variations of ϕ from one experiment to another, and to certain apparent absences of complete randomness in the actual occurrences of ξ in successive trials.

independent quantities equals the product of their means. Then by postulate (2),

$$1 + \alpha[x] + \tfrac{1}{2}\alpha^2[x^2] + \ldots + \frac{1}{l!}\alpha^l[x^l] + \ldots$$
$$= \prod_{q=1}^{q=m}\left(1 + \tfrac{1}{2}\alpha^2[\xi_q{}^2] + \ldots + \frac{1}{l!}\alpha^l[\xi_q{}^l] + \ldots\right),$$

since $[\xi_q]$ is zero.

Write M_l for $[x^l]$ and ${}_q\mu_l$ for $[\xi_q{}^l]$, that is for the lth moment of the qth elemental curve. Take logarithms of both sides, so as to replace multiplication by addition. We have

$$\log\left(1 + \alpha \,.\, M_1 + \tfrac{1}{2}\alpha^2 M_2 + \ldots + \frac{1}{l!}\alpha^l M_l + \ldots\right)$$
$$= \sum_{q=1}^{q=m} \log\left(1 + \tfrac{1}{2}\alpha^2 \,.\, {}_q\mu_2 + \ldots + \frac{1}{l!}\alpha^l \,.\, {}_q\mu_l + \ldots\right).$$

Expand and equate the coefficients of α on the two sides, likewise for α^2:

$M_1 = 0$, $M_2 = \Sigma_q\mu_2$, or if S and σ_q are written for the standard deviations, $S^2 = \Sigma\sigma_q{}^2$.

Let $m\mu_2 = \Sigma_q\mu_2$, so that μ_2 is the average of the second moments of the elemental curves, and we obtain $M_2 = m\mu_2$, $S = \sqrt{m}\sqrt{\mu_2}$.

Now S is evidently a finite quantity in the frequency curve of x, which we seek, and therefore $\sqrt{\mu_2}$ is of order $m^{-\frac{1}{2}}$, and $\sigma_1 \ldots \sigma_q \ldots \sigma_m$ are also of order $m^{-\frac{1}{2}}$, since the elemental curves by the third postulate are similar in scale. Generally, the ratio $1 : m^{\frac{1}{2}}$ is the ratio of the typical elemental deviation to the typical deviation of x.

Consider m to be large and arrange the expression in ascending powers of $m^{-\frac{1}{2}}$. The larger is m, the more rapid the convergence.

Now ${}_q\mu_l/\sigma_q{}^l$ is finite by the third postulate, so that ${}_q\mu_l$ is of order $m^{-\frac{l}{2}}$ for all values of q; hence in the expression $1 + \tfrac{1}{2}\alpha^2 \,.\, {}_q\mu_2 + \ldots + \frac{1}{l!}\alpha^l \,.\, {}_q\mu_l + \ldots$ the coefficient of α^l is of order $m^{-\frac{l}{2}}$, and the same statement holds for its logarithmic expansion.

$$\log\left(1+\frac{1}{2}\alpha^2\,.\,{}_q\mu_2+\ldots\right)=\frac{1}{2}\alpha^2\,.\,{}_q\mu_2+\frac{1}{3!}\alpha^3\,.\,{}_q\mu_3$$
$$+\ldots-\frac{1}{2}\left(\frac{1}{2}\alpha^2\,.\,{}_q\mu_2+\frac{1}{3!}\alpha^3\,.\,{}_q\mu_3+\ldots\right)^2$$
$$+\tfrac{1}{3}(\tfrac{1}{2}\alpha^2\,.\,{}_q\mu_2+\ldots)^3+\ldots$$
$$=\frac{1}{2}\alpha^2\,.\,{}_q\mu_2+\frac{{}_qk_1}{3!}\alpha^3+\frac{{}_qk_2}{4!}\alpha^4+\ldots+\frac{{}_qk_l}{l!}\alpha^l+\ldots$$

where ${}_qk_l$ is of order $m^{-\frac{l}{2}}$ and the ${}_qk$'s may be worked out in terms of the moments μ.

Thus ${}_qk_1={}_q\mu_3$, which is zero if the qth elemental curve is symmetrical.

${}_qk_2={}_q\mu_4-3{}_q\mu_2{}^2$, which is zero if the qth elemental curve is normal.

It is evident that every k with an odd suffix is zero if the elemental curve is normal and therefore symmetrical, and it can be shown that the k's with even suffixes are zero under the same condition.* Thus the ${}_qk$'s depend on the abnormality of the elemental curves.

Returning to the main equation and removing the logarithm on the left-hand side, we have

$$1+\frac{1}{2}\alpha^2\,.\,M_2+\ldots+\frac{1}{l!}\alpha^l\,.\,M_l+\ldots=e^{\Sigma\left(\frac{1}{2}\alpha^2\,.\,{}_q\mu_2+\frac{1}{3!}k_1\alpha^3+\ldots\right)}$$
$$=e^{\frac{1}{2}M_2\alpha^2}\,.\,e^{\frac{1}{3!}k_1\alpha^3+\frac{1}{4!}k^2\alpha^4+\frac{1}{5!}k_3\alpha^5+\ldots}$$

where

$$k_1=\sum_{q=1}^{q=m}{}_qk_1=\Sigma_qm_3\,;\quad k_2=\sum_{q=1}^{q=m}{}_qk_2=\Sigma({}_q\mu_4-3{}_q\mu_2{}^2)\,;$$
$$\ldots\ k_l=\Sigma_qk_l;\ \ldots$$

Σ_qk_l is the sum of m terms each of order $m^{-\frac{l}{2}}$, and is

* In the normal curve ${}_q\mu_{2l+1}=0$ and ${}_q\mu_{2l}=\dfrac{2l!}{2^l\,.\,l!}\,.\,{}_q\sigma^{2l}$.

$$\therefore\ \log(1+\tfrac{1}{2}a^2\,.\,{}_q\mu_2+\ldots)=\log\left(1+\frac{1}{2}a^2\,.\,{}_q\sigma^2+\frac{1}{4!}\cdot\frac{4!}{2^2 2!}\cdot{}_q\sigma^4a^4+\ldots\right.$$
$$\left.+\frac{1}{2l!}\cdot\frac{(2l)!}{2^l\,.\,l!}\cdot{}_q\sigma^{2l}\,.\,a^{2l}+\ldots\right)=\log\left(e^{\frac{1}{2}a^2\,.\,{}_q\sigma^2}\right)=\tfrac{1}{2}a^2\,.\,{}_q\mu_2,$$

and it follows that ${}_qk_1$, ${}_qk_2$ . . . are each zero.

This statement is implied, but not emphasised, in Edgeworth's papers.

therefore of order $m^{-\frac{l}{2}+1}$. Then k_1 is of order $m^{-\frac{1}{2}}$, k_2 of order m^{-1}, etc.

By expanding and equating coefficients we may obtain the values of as many M's in terms of k's as we please.

$$M_3 = k_1 = \Sigma_q\mu_3.$$
$$M_4 = k_2 + 4!\,.\,\tfrac{1}{2}(\tfrac{1}{2}M_2{}^2). \quad M_4 - 3M_2{}^2 = k_2 = \Sigma({}_q\mu_4 - 3_q\mu_2{}^2).$$
$$M_5 = k_3 + 10M_2k_1.$$
etc., etc.

In particular, if all the elemental curves have the same standard deviation and moments, σ, μ_3, μ_4 . . ., $\dfrac{M_3}{S^3} = \dfrac{m\mu_3}{m^{\frac{3}{2}}\sigma^3} = \dfrac{1}{\sqrt{m}}\left(\dfrac{\mu_3}{\sigma^3}\right)$, which gives the relation between the asymmetry of the resulting curve and that of the elemental curve, and

$$\frac{M_4}{S^4} - 3 = \frac{m(\mu_4 - 3\mu_2{}^2)}{m^2\sigma^4} = \frac{1}{m}\left(\frac{\mu_4}{\sigma^4} - 3\right),$$

which gives the relation between the "kurtosis" of the two curves.

The More General Formula

A further step is necessary to obtain a general expression for the frequency or x. Consider the expression, or "generating function,"

$$z = e^{-\frac{1}{3!}k_1D^3 + \frac{1}{4!}k_2D^4 - \ldots + (-1)^t\frac{1}{(t+2)!}k_tD^{t+2} + \ldots} \, . \, z_0$$

where $z_0 = \dfrac{1}{c\sqrt{\pi}}e^{-\frac{x^2}{c^2}}$, $c^2 = 2M_2$, and D stands for differentiation with respect to x.

Now $\int_{-\infty}^{\infty} x^t \, . \, D^{t-2r} \, . \, z_0dx$, where r is any integer up to $\dfrac{t}{2}$,

$$= \Big[x^t \, . \, D^{t-2r-1} \, . \, z_0\Big]_{-\infty}^{\infty} - t \, . \int_{-\infty}^{\infty} x^{t-1}D^{t-2r-1} \, . \, z_0dx$$
$$= - t \, . \int_{-\infty}^{\infty} x^{t-1} \, . \, D^{t-2r-1} \, . \, z_0dx$$
$$= \text{(similarly) } t(t-1)\int_{-\infty}^{\infty} x^{t-2} \, . \, D^{t-2r-2} \, . \, z_0dx.$$

Carry on the process till the index of D is 0 :

$$\int_{-\infty}^{\infty} x^t \,.\, D^{t-2r} \,.\, z_0 dx = (-1)^{t-2r} \,.\, \frac{t!}{(2r)!}\int_{-\infty}^{\infty} x^{t-(t-2r)} \,.\, z_0 dx$$

$$= (-1)^{t-2r} \,.\, \frac{t!}{(2r)!} \,.\, \frac{(2r)!}{2^r \,.\, r!} M_2^r$$

$$= (-1)^{t-2r} \,.\, \frac{t!}{2^r \,.\, r!} M_2^r.$$

Take the case when t is even. A simple counting of signs shows that the result is true also when t is odd. Now

$$\int_{-\infty}^{\infty} x^t \,.\, z dx =$$

$$\int_{-\infty}^{\infty}\left\{1 + \ldots \frac{1}{l!}\left(-\frac{1}{3!}k_1 D^3 + \frac{1}{4!}k_2 D^4 - \ldots\right)^l + \ldots\right\} x^t z_0 dx.$$

The coefficient of D^{t-2r} is, by the multinomial theorem, the sum of all such expressions as

$$Q = \frac{1}{l!}\left(\frac{k_1}{3!}\right)^{a_1}\left(\frac{k_2}{4!}\right)^{a_2} \ldots \times \frac{l!}{a_1! a_2! \ldots},$$

where $a_1, a_2 \ldots$ have all positive integral values consistent with $a_1 + a_2 + \ldots = l$, and $3a_1 + 4a_2 + 5a_3 + \ldots = t - 2r$, and l is given successively all possible values.

We have therefore the result that $\int_{-\infty}^{\infty} x^t z dx$, the tth moment (t even) of the curve given by the generating function is the sum of the terms $Q \,.\, \frac{t!}{2^r \,.\, r!} M_2^r$ for all permissible values of Q and r.

Now this is precisely the expression obtained for the coefficient of α^t; for from the equation above (p. 42), $\frac{1}{t!} M_t =$ coefficient of α in

$$\left\{\ldots \frac{1}{l!}\left(\frac{1}{2}M_2\alpha^2 + \frac{1}{3!}k_1\alpha^3 + \frac{1}{4!}k_2\alpha^4 + \ldots\right)^l \ldots\right\},$$

that is the sum of all such expressions as

$$\frac{1}{l!}\left(\frac{1}{2}M_2\right)^r\left(\frac{1}{3!}k_1\right)^{a_1}\left(\frac{1}{4!}k_2\right)^{a_2} \ldots \,.\, \frac{l!}{r! a_1! a_2! \ldots} = Q(\tfrac{1}{2}M_2)^r / r!$$

where $r + a_1 + a_2 + \ldots = l + r$, $2r + 3a_1 + 4a_2 + \ldots = t$, and $l, a_1, a_2 \ldots$ and r have the same ranges of values as before.

$$\therefore\ M_t = \text{sum of the terms } Q\frac{(\frac{1}{2}M_2)^r t!}{r!} = \int_{-\infty}^{\infty} x^t z dx.$$

Thus the curve $z = e^{-\frac{1}{3!}k_1D^3 + \frac{1}{4!}k_2D^4 - \dots} . z_0$ has the same moments up to the tth as are given by the variation of ${}_px$, where t may be taken as great as we please, and it therefore expresses the frequency curve of ${}_px$.

Properties of the Curve

The expression may be worked out at any length, but it is found in practice that only the first, second and third approximations are needed. The first is obtained by neglecting terms in $\frac{1}{\sqrt{m}}$, that is by taking k_1, k_2 . . . as zero, and is the familiar Gaussian law. The second and third are obtained by neglecting $\frac{1}{m}$, *i.e.* taking k_2, k_3 . . . as zero, and $\frac{1}{m^{\frac{3}{2}}}$, *i.e.* taking k_3, k_4 . . . as zero, respectively.

The third approximation is

$$y = \left(1 - \frac{1}{3!}k_1D^3 + \frac{1}{4!}k_2D^4 + \frac{1}{2(3!)^2}k_1{}^2D^6\right)\frac{1}{c\sqrt{\pi}}e^{-\frac{x^2}{c^2}},$$

since $k_1{}^3$, k_1k_2, . . . are of order $m^{-\frac{3}{2}}$ which is neglected,

$$= \frac{1}{c\sqrt{\pi}}e^{-\frac{x^2}{c^2}}\left\{1 - j\left(2\frac{x}{c} - \frac{4}{3}\cdot\frac{x^3}{c^3}\right) + i\left(\frac{1}{2} - \frac{2x^2}{c^2} + \frac{2}{3}\cdot\frac{x^4}{c^4}\right)\right.$$
$$\left. + j^2\left(-\frac{5}{3} + 10\frac{x^2}{c^2} - \frac{20}{3}\ \frac{x^4}{c^4} + \frac{8}{9}\cdot\frac{x^6}{c^6}\right)\right\}$$

where $c^2 = 2M_2$ is the modulus-squared of the curve, as may be verified by integrating $\int_{-\infty}^{\infty} 2x^2y dx$. The origin is the mean of the curve, as may be verified by showing that $\int_{-\infty}^{\infty} xydx = 0$, and the area is unity.

Here j is written for $\frac{k_1}{c^3} = \frac{M_3}{c^3}$, and $j^2 = \frac{1}{8}\beta_1{}^2$, in Prof. Pearson's notation; and i is written for

$$\frac{k_2}{c^4} = \frac{M_4 - 3M_2{}^2}{4M_2{}^2} = \tfrac{1}{4}(\beta_2 - 3)$$

in Prof. Pearson's notation.

Write $\frac{x}{c} = \tau$, thus expressing x in terms of the modulus, and write $F(\tau)$ for $\int_0^\tau \frac{1}{\sqrt{\pi}} e^{-\tau^2} d\tau$ and $f(\tau)$ for $\frac{1}{\sqrt{\pi}} e^{-\tau^2}$, and integrate:

$$\int_0^x ydx = F(\tau) - \frac{j}{3\sqrt{\pi}} + \frac{j}{3}(1 - 2\tau^2)f(\tau) +$$
$$\left\{i\left(\frac{\tau}{2} - \frac{\tau^3}{3}\right) + j^2\left(-\frac{5}{3}\tau + \frac{20}{9}\tau^3 - \frac{4}{9}\tau^5\right)\right\}f(\tau).$$

In any area $\int_{x_1}^{x_2} ydx$ the constant $\frac{j}{3\sqrt{\pi}}$ drops out.

The second approximation is obtained by neglecting terms involving i and j^2. In this case $\int_{-x}^{x} ydx = 2F(\tau) = 2F\left(\frac{x}{c}\right)$.

If i and j^2 are neglected,

$$\frac{1}{y} \cdot \frac{dy}{dx} = \frac{d}{dx}(\log y) = -\frac{2x}{c^2} - j\left(\frac{2}{c} - \frac{4x^2}{c^3}\right) = -\frac{2}{c^2}x'\left(1 - \frac{2}{c}x'j\right)$$
$$= -\frac{x'}{\frac{c^2}{2} + jcx'}$$

where $x' = x + jc$, and j^2 is neglected throughout. This is Prof. Pearson's equation from the hypergeometric series in the case where his b_2 is negligible.

Returning to the third approximation, we will find the mode and the median. For the former,

$$0 = \frac{d}{dx}(\log y) = -\frac{2x}{c^2} - j\left(\frac{2}{c} - \frac{4x^2}{c^3}\right) + i\left(-\frac{4x}{c^2} + \frac{8x^3}{3c^4}\right)$$
$$+ j^2\left(\frac{16x}{c^2} - 16\frac{x^3}{c^4}\right)$$

when ij, i^2, etc., are neglected. The solution of this equation to the same order of approximation is $x = -jc$.

For the median we must find x so that

$$\int_0^x ydx + \frac{1}{2} = \int_0^\infty ydx = \frac{1}{2} - \frac{j}{3\sqrt{\pi}}.$$

x is evidently of the order j. Expand $\int_0^x ydx$ by Taylor's Theorem:

$$0 + x\left(\frac{1}{c\sqrt{\pi}}\right)\left(1 + \frac{i}{2}\right) + \frac{x^2}{2} \cdot \frac{1}{c\sqrt{\pi}}\left(-\frac{2j}{c}\right) + \ldots + \frac{1}{2} = \frac{1}{2} - \frac{j}{3\sqrt{\pi}}.$$

When j^3, ij and c are neglected the solution is $x = -\frac{1}{3}jc$. The area on the base median-to-mean is $-\dfrac{j}{3\sqrt{\pi}}$.

In the particular case when every elemental frequency group contains qN zeros and pN units, which expresses the simple case that the chance of adding 1 to the magnitude (success) is p, and the chance of adding nothing (failure) is q, $(p + q = 1)$, the constants for the elements are

$$\sigma = \sqrt{pq}, \quad \mu_3 = \frac{pq^3 - qp^3}{p + q} = pq(q - p),$$

$$\mu_4 = \frac{pq^4 + qp^4}{p + q} = pq(1 - 3pq)$$

$\therefore$ from $p.$ 41 $M_2 = mpq$, $c^2 = 2mpq$, $k_1 = mpq(q - p)$, $j = \dfrac{q - p}{2\sqrt{2mpq}}$, $k_2 = mpq(1 - 6pq)$, $i = \dfrac{1 - 6pq}{4mpq}$.

The frequency curve for an average, $\dfrac{1}{m} \cdot x$, has for its modulus $\dfrac{\sqrt{2M_2}}{m} = \dfrac{\sigma\sqrt{2}}{\sqrt{m}}$ if σ^2 is the average of $\sigma_1^2, \ldots \sigma_q^2, \ldots \sigma_m^2$. j and i are the same as for the sum, and the frequency curve is that given for the sum with the reduced modulus $\dfrac{c}{\sqrt{m}}$.

The modulus for a weighted average, where the elements $\xi_1 \ldots \xi_q \ldots$ are multiplied by $w_1 \ldots w_2 \ldots$, is $\dfrac{\sqrt{2\Sigma w_q^2 \cdot \sigma_q^2}}{\Sigma w_q}$.

Two or More Dimensions

The extension to two dimensions (given very briefly in (47), pp. 116–19, but amplified in (58), *J.S.S.*, 1917, pp. 266 *seq.*) follows the same general lines. The analysis is simplified by referring all the variables to the principal axes of the correlation surface.

We have now m elemental frequency surfaces, in which the probability of the joint occurrence of deviations from the double average may be written $\phi_q(\xi_q\eta_q)$ for the qth surface. ϕ may be of any form, except that it is limited as before for the moments of each variable separately, and also by the condition that $\mu_{tt'} \div \sigma_1^t\sigma_2^{t'}$ shall differ from what it would be in a normal surface by a finite quantity, where σ_1, σ_2 are the standard deviations in

ξ and η, and $\mu_{tt'} = [\xi^t \eta^{t'}]$, the limitation applying to all values of t and t'.

We have to find the moments and frequency surface for the occurrence of pairs of values ${}_p x$, ${}_p y$, where

$$ {}_p x = \sum_{q=1}^{q=m} {}_p\xi_q, \text{ and } {}_p y = \sum_{q=1}^{q=m} {}_p\eta_q. $$

Write $m_{tt'} = [x^t y^{t'}]$ for all values of t and t' including zero. Treat x and y as measured on rectangular axes. Now transfer to rectangular axes making the angle θ with OX, where

$$ \tan 2\theta = \frac{2m_{11}}{m_{20} - m_{02}}. $$

Write ${}_p X = {}_p x \cos\theta + {}_p y \sin\theta$, ${}_p Y = - {}_p x \sin\theta + {}_p y \cos\theta$, and write $M_{tt} = [X^t Y^{t'}]$, $K = M_{20} = [X^2]$, $K' = M_{02} = [Y^2]$.

Then $M_{11} = 0$.

Finally write

$$ {}_p u_q = {}_p\xi_q \cos\theta + {}_p\eta_q \sin\theta,\ {}_p v_q = - {}_p\xi_q \sin\theta + {}_p\eta_q \cos\theta. $$

Then $$ {}_p X = \sum_{q=1}^{q=m} {}_p u_q,\ {}_p Y = \sum_{q=1}^{q=m} {}_p v_q. $$

$$ K = M_{20} = m_{20}\cos^2\theta + m_{11}\sin 2\theta + m_{02}\sin^2\theta. $$
$$ K' = M_{02} = m_{20}\sin^2\theta - m_{11}\sin 2\theta + m_{02}\cos^2\theta. $$

r, the correlation coefficient between x and y, $= \dfrac{m_{11}}{\sqrt{m_{20}m_{02}}}$,

where $m_{20} = \sum_{q=1}^{q=m} [\xi_q^2]$, $m_{02} = \sum_{q=1}^{q=m} [\eta_q^2]$, $m_{11} = \sum_{q=1}^{q=m} [\xi_q \eta_q]$.

Let α, β be any small constant quantities to be used for collecting coefficients.

Expand both sides and take the mean for all values of p, remembering that ${}_p u_q$ is independent of ${}_p u_{q'}$, where q, q' refer to any pair of elemental surfaces.

$$ 1 + [\alpha X + \beta Y] + \ldots + \frac{1}{l!}[(\alpha X + \beta Y)^l] + \ldots $$
$$ = \Pi\{1 + \tfrac{1}{2}(\alpha . u_q + \beta . v_q)^2 + \ldots\}, $$

since $[u_q] = [\xi_q]\cos\theta + [\eta_q]\sin\theta = 0$ and $[v_q]$ similarly $= 0$,

$$ = e^{\sum_{q=1}^{q=m} \log\left\{1 + \frac{1}{2}[(\alpha u_q + \beta v_q)^2] + \frac{1}{3!}[(\alpha u_q + \beta v_q)^3] + \ldots\right\}}. $$

On the left hand the coefficient of $\alpha^t\beta^{t'}$, where t, t' are any integers or zero and their sum is l, is

$$\frac{1}{l!}[X^tY^{t'}]\frac{l!}{t!t'!}=\frac{M_{tt'}}{t!t'!}.$$

Expanding and equating the first few coefficients, we get

$$M_{10}=0=M_{01};\ K=M_{20}=\sum_{q=1}^{q=m}[u_q]^2;\ K'=M_{02}=\sum_{q=1}^{q=m}[v_q]^2;$$
$$\Sigma[u_qv_q]=M_{11}=0;\ M_{30}=\Sigma[u_q^3];\ M_{21}=3\Sigma[u_q^2v_q];$$
$$M_{12}=3\Sigma[u_qv_q^2];\ M_{03}=\Sigma[v_q^3].$$

The coefficient of $\alpha^t\beta^{t'}$ in the index is the sum of m terms each of order $[u_q^tv_q^{t'}]$, that is of order $m\sigma_1^t\sigma_2^{t'}$, where σ_1, σ_2 are typical standard deviations of the elemental curve. But σ_1, σ_2 are of order $Km^{-\frac{1}{2}}$, $K'm^{-\frac{1}{2}}$; therefore the coefficient of $\alpha^t\beta^{t'}$ is of order $m^{-\frac{t+t'}{2}+1}$.

In the sequel the analysis is not pursued beyond the second approximation, terms of order m^{-1} or higher being neglected. To this order we have

$$1+\tfrac{1}{2}\alpha^2K+\tfrac{1}{2}\beta^2K'+\ .\ .\ .\ +\frac{\alpha^t\beta^t\,.\,M_{tt'}}{t!t'!}+\ .\ .\ .$$
$$=e^{\frac{1}{2}\alpha^2K+\frac{1}{2}\beta^2K'+\frac{1}{6}(\alpha^3M_{30}+3\alpha^2\beta M_{21}+3\alpha\beta^2M_{12}+\beta^3M_{03})}.$$

Write $$Z_0=\frac{1}{\pi\sqrt{KK'}}e^{-\frac{1}{2}\left(\frac{X^2}{K}+\frac{Y^2}{K'}\right)},$$

and $$Z=e^{-\frac{1}{6}(M_{30}D_X^3+3M_{21}D_X^2D_Y+3M_{12}D_XD_Y^2+M_{03}D_Y^3)}\,.\,Z_0,$$

where D_X, D_Y signify partial differentiation with respect to X and Y. Then, as in the case of one dimension (p. 43–44 above),

$\int_{-\infty}^{\infty}\int_{-\infty}^{\infty}X^tY^{t'}ZdXdY$ can be identified with the coefficient of $\frac{\alpha^t\beta^{t'}}{t!t'!}$ in the expansion of $e^{\frac{1}{2}\alpha^2K+\text{etc.}}$, above.*

Therefore the moments of Z of all orders are the corresponding moments of (X, Y), and Z gives the frequency surface for X, Y, when terms involving $\frac{1}{m}$ are neglected.

* Not only in the terms here included, but for approximations carried to any power of $m^{-\frac{1}{2}}$.

Since M_{30}^2, M_{21}^2, . . . are negligible we may write the equation

$$Z = \left\{1 - \frac{1}{6}(M_{30}Dx^3 + \ . \ . \ .)\right\}\frac{1}{2\pi\sqrt{KK'}}e^{-\frac{1}{2}\left(\frac{X^2}{K} + \frac{Y^2}{K'}\right)}$$

$$= \Big[1 - \frac{1}{6}\Big\{M_{30}\left(\frac{3X}{K^2} - \frac{X^3}{K^3}\right) + 3M_{21}\left(\frac{Y}{KK'} - \frac{X^2Y}{K^2K'}\right)$$

$$+ 3M_{12}\left(\frac{X}{KK'} - \frac{XY^2}{KK'^2}\right) + M_{03}\left(\frac{3Y}{K'^2} - \frac{Y^3}{K'^3}\right)\Big\}\Big]$$

$$\frac{1}{2\pi\sqrt{KK'}}e^{-\frac{1}{2}\left(\frac{X^2}{K} + \frac{Y^2}{K'}\right)}.$$

It remains to transfer this equation to the original axes.*

We have $X = x \cos \theta + y \sin \theta$. $Y = - x \sin \theta + y \cos \theta$.

$$\frac{\sin 2\theta}{2m_{11}} = \frac{\cos 2\theta}{m_{20} - m_{02}} = \frac{1}{b}, \text{ where } b^2 = (m_{20} - m_{02})^2 + 4m_{11}^2,$$

and where $m_{11} = [xy]$, etc.

$$K = M_{20} = [X^2] = m_{20} \cos^2 \theta + m_{11} \sin 2\theta + m_{02} \sin^2 \theta,$$
$$K' = m_{20} \sin^2 \theta - m_{11} \sin 2\theta + m_{20} \cos^2 \theta,$$
$$\therefore\ 2K = m_{20} + m_{02} + b,\ 2K' = m_{20} + m_{02} - b,$$
$$K + K' = m_{20} + m_{02},\ K - K' = b,\ KK' = m_{20}m_{02} - m_{11}^2,$$
$$K \sin^2 \theta + K' \cos^2 \theta = m_{02},\ K \cos^2 \theta + K' \sin^2 \theta = m_{20}.$$

r, the coefficient of correlation between x and y,

$$= \frac{m_{11}}{\sqrt{m_{20} \cdot m_{02}}}.$$

Write $\rho^2 = 1 - r^2$.

If c_1, c_2 are the moduli for x and y, $c_1^2 = 2m_{20}$, $c_2^2 = 2m_{02}$.

After a simple substitution we obtain

$$z_0 = \frac{1}{2\pi\sqrt{KK'}}e^{-\frac{1}{2}\left(\frac{X^2}{K} + \frac{Y^2}{K'}\right)} = \frac{1}{\pi c_1 c_2 \rho}e^{-\left(\frac{x^2}{c_1^2} + \frac{y^2}{c_2^2} - \frac{2rxy}{c_1c_2}\right)/\rho^2},$$

the equation for the normal surface.

* Edgeworth gives a difficult short cut for the general solution, based on the theory of emanants ((47), pp. 118–19), by which it is shown that the expression

$$M_{30}\frac{d^3Z_0}{dX^3} + 3M_{21} \cdot \frac{d^2Z_0}{dX^2} \cdot \frac{dZ_0}{dY} + \text{etc.},$$

can be replaced by

$$m_{30}\frac{d^3z_0}{dx^3} + 3m_{21}\frac{d^2z_0}{dx^2} \cdot \frac{dz_0}{dy} + \text{etc.}$$

It seemed better to write the solution for the second approximation more directly.

Further,

$$M_{30}=[X^3]=m_{30}\cos^3\theta+3m_{21}\cos^2\theta\sin\theta$$
$$+3m_{12}\cos\theta\sin^2\theta+m_{03}\sin^3\theta$$
$$M_{21}=[X^2Y]=-m_{30}\cos^2\theta\sin\theta+m_{21}\cos\theta\,(1-3\sin^2\theta)$$
$$+m_{12}\sin\theta\,(3\cos^2\theta-1)+m_{03}\sin^2\theta\cos\theta$$

with similar expressions for M_{12} and M_{03}.

Write these values for M_{30}, etc., in the brackets above and collect terms in m_{30}, etc. It will be found that the coefficients of m_{30}, m_{21}, m_{12}, m_{03} are expressible in terms of

$$\frac{X}{K}\cos\theta-\frac{Y}{K'}\sin\theta=\frac{m_{02}x-m_{11}y}{m_{20}m_{02}-m_{11}{}^2}$$

and

$$\frac{X}{K}\sin\theta+\frac{Y}{K'}\cos\theta=\frac{m_{20}y-m_{11}x}{m_{20}m_{02}-m_{11}{}^2}.$$

It is found convenient to use the quantities $q=\left(\frac{x}{c_1}-r\frac{y}{c_2}\right)\Big/\rho$ and $q'=\left(\frac{y}{c_2}-r\frac{x}{c_1}\right)\Big/\rho$; it is easily seen that

$$\frac{X}{K}\cos\theta-\frac{Y}{K'}\sin\theta=\frac{2}{c_1\rho}q \quad\text{and}\quad \frac{X}{K}\sin\theta+\frac{Y}{K'}\cos\theta=\frac{2}{c_2\rho}q'.$$

The whole expression then becomes

$$z=\left[1-\frac{1}{\rho^3}\left\{\frac{m_{30}}{c_1{}^3}\left(2q-\frac{4}{3}q^3\right)+\frac{m_{21}}{c_1{}^2c_2}(2q'-4rq-4q^2q')\right.\right.$$
$$\left.\left.+\frac{m_{12}}{c_1c_2{}^2}(2q-4rq'-4qq'^2)+\frac{m_{03}}{c_2{}^3}\left(2q'-\frac{4}{3}q'^3\right)\right\}\right].z_0,$$

where

$$z_0=\frac{1}{\pi c_1c_2\rho}e^{-\left(\frac{x^2}{c_1{}^2}+\frac{y^2}{c_2{}^2}-\frac{2rxy}{c_1c_2}\right)\Big/\rho^2}=\frac{1}{\pi c_1c_2\rho}e^{-q^2}.e^{\frac{y^2}{c_2{}^2}}$$
$$=\frac{1}{\pi c_1c_2\rho}e^{-q'^2}.e^{\frac{x^2}{c_1{}^2}}.$$

The *curve of regression*, taken as the locus of the modes of arrays in x, is obtained by equating $\frac{\partial}{\partial y}.z$ to zero.

When we take the first approximation only we have

$$0=\frac{1}{2}\frac{\partial}{\partial y}\log z_0=\frac{1}{c_2\rho}\left(\frac{y}{c_2}-r\frac{x}{c_1}\right),$$

that is, $q'=0$.

Deviations from this due to the inclusion of m_{30}, etc., may be regarded as depending on small changes in q' of the order $m^{-\frac{1}{2}}$; and $m_{30}q'$, etc., can be neglected after differentiation.

Now $q = \frac{x}{c_1}\rho - rq'$, and after differentiation q can be replaced by ρ.

Also $\frac{\partial q}{\partial y} = -\frac{r}{\rho c_2}, \frac{\partial q'}{\partial y} = \frac{1}{\rho c_2}$. Hence to the same order of approximation $0 = \frac{1}{2}\frac{\partial}{\partial y}(\log z)$ gives

$$0 = -q'\frac{\partial q'}{\partial y} - \frac{1}{\rho^3}\left[\frac{m_{30}}{c_1^{\,3}}\left(1 - 2\frac{x^2}{c_1^{\,2}}\rho^2\right)\left(-\frac{r}{\rho c_2}\right) + \frac{m_{21}}{c_1^{\,2}c_2}\left\{(-2r)\left(-\frac{r}{\rho c_2}\right) + \frac{1}{\rho c_2} - 2\frac{x^2}{c_1^{\,2}}\rho^2 \cdot \frac{1}{\rho c_2}\right\} + \frac{m_{12}}{c_1 c_2^{\,2}}\left(-\frac{r}{\rho c_2} - \frac{2r}{\rho c_2}\right) + \frac{m_{03}}{c_2^{\,3}} \cdot \frac{1}{\rho c_2}\right]$$

which leads to

$$y = r\frac{c_2}{c_1}x + \frac{c_2}{\rho^2}\left\{\frac{m_{30}}{c_1^{\,3}}r - \frac{m_{21}}{c_1^{\,2}c_2}(1 + 2r^2) + \frac{m_{12}}{c_1 c_2^{\,2}} \cdot 3r * - \frac{m_{03}}{c_2^{\,3}}\right\} - 2c_2\left(r\frac{m_{30}}{c_1^{\,3}} - \frac{m_{21}}{c_1^{\,2}c_2}\right)\frac{x^2}{c_1^{\,2}}.$$

The *equation of the section by* $x = x_1$ may be found without loss of generality in the case where $c_1 = c_2 = 1$, in terms of q' and x, where $q = x\rho - rq'$, and $y = \rho q' + rx$.

Then

$$Z = \left[\begin{array}{l} 1 - \frac{1}{\rho^3}\left\{m_{30}\left(2 \cdot \overline{x\rho - rq'} - \frac{4}{3} \cdot \overline{x\rho - rq'}^{\,3}\right)\right. \\ \quad + m_{21}(2q' - 4r \cdot \overline{x\rho - rq'} - 4q' \cdot \overline{x\rho - rq'}^{\,2}) \\ \quad + m_{12}(2 \cdot \overline{x\rho - rq'} - 4rq' - 4q'^2 \cdot \overline{x\rho - rq'}) \\ \quad \left. + m_{03}\left(2q' - \frac{4}{3}q'^3\right)\right\} \end{array}\right] \frac{e^{-q'^2}}{\sqrt{\pi}} \cdot \frac{e^{-x^2}}{\rho\sqrt{\pi}}.$$

The *area* of this section $= \int_{-\infty}^{\infty} z dy$, x constant, $= \int_{-\infty}^{\infty} \rho z dq'$

$$= \frac{e^{-x^2}}{\sqrt{\pi}}\left\{1 - \frac{1}{\rho^3}m_{30}\left(2x\rho - \frac{4}{3}x^3\rho^3 - 2x\rho r^2\right)\right\},$$

all other terms disappearing on integration,

$$= \frac{e^{-x^2}}{\sqrt{\pi}}\left\{1 - m_{30}\left(2x - \frac{4}{3}x^3\right)\right\}.$$

* Edgeworth gives r instead of $3r$ here by error.

The *mean* of the section is

$$\frac{\int_{-\infty}^{\infty} yzdy}{\int_{-\infty}^{\infty} zdy} = \frac{\int_{-\infty}^{\infty} (rx + \rho q')zdy}{\int_{-\infty}^{\infty} zdy} = rx + \frac{\int_{-\infty}^{\infty} \rho^2 q' zdq'}{\int_{-\infty}^{\infty} zdy}.$$

The numerator of the second term is

$$\frac{\rho e^{-x^2}}{\sqrt{\pi}}\left\{-\frac{1}{\rho^3}(-r + 2x^2\rho^2 r + r^3)m_{30} + (1 + 2r^2 - 2x^2\rho^2 - 3r^2)m_{21}\right\}$$

(the other terms disappearing),

$$= \frac{e^{-x^2}}{\sqrt{\pi}}(m_{30}\, r - m_{21})(1 - 2x^2).$$

The mean of the x-array is therefore

$$rx + \frac{(m_{30}r - m_{21})(1 - 2x^2)}{1 - m_{30}(2x - \frac{4}{3}x^3)} = rx + (m_{30}r - m_{21})(1 - 2x^2),*$$

correct to the order of approximation, terms in $\frac{1}{m}$ being neglected.

The locus of the mean (now restoring c_1, c_2) is

$$y = r\frac{c_2}{c_1}x + c_2\left(\frac{m_{30}}{c_1^3}r - \frac{m_{21}}{c_1^2 c_2}\right)\left(1 - 2\frac{x^2}{c_1^2}\right).$$

The *volume* of the surface standing on any small area may be found by regarding the equation as $z = z_0\ (1 - w)$, where w contains the terms m_{30}, etc., then finding the part of the volume due to the first approximation by the ordinary method, and subtracting the ordinate wz_0 multiplied by the area (expressed in the appropriate units). This is especially applicable to discrete observations ((58), *J.S.S.*, 1917, p. 268). Direct and complete integration is, of course, possible, but arduous.

A simpler method is as follows. From the beginning of this section (p. 47) write $x = c_1 x'$ and $y = c_2 y'$, so that the moduli of x' and y' are unity. For these variables $2m_{20} = c_1^2 = 1$, $2m_{02} = c_2^2 = 1$. Then $\tan 2\theta$ is infinite, and

$$\cos\theta = \sin\theta = \frac{1}{\sqrt{2}}.\quad r = 2m_{11}.\quad 2K = 1 + r,\ 2K' = 1 - r.$$

* Edgeworth inserts an erroneous factor $\frac{1}{\rho}$ in the corrective term.

Now write $X = X'\sqrt{1+r} = X'\sqrt{2K},$

and $Y = Y'\sqrt{1-r} = Y'\sqrt{2K'},$

then $$Z_0 = \frac{1}{\pi\rho^2}e^{-(X'^2+Y'^2)},$$

and $$Z = Z_0 - \frac{1}{6}\left\{\frac{M_{30}}{K^{\frac{3}{2}}}\sqrt{2}(3X' - 2X'^3) + 3\frac{M_{21}}{KK'^{\frac{1}{2}}}\sqrt{2}(Y' - 2X'^2Y') + 3\frac{M_{12}}{K^{\frac{1}{2}}K'}\sqrt{2}(X' - 2X'Y'^2) + \frac{M_{03}}{K'^{\frac{3}{2}}}\sqrt{2}(3Y' - 2Y'^3)\right\}Z_0.$$

Write K_{30} for $[X'^3]$, K_{21} for $[X'^2Y']$, etc.

By integration we obtain

$$\int_0^{X'}\int_0^{Y'} ZdX'dY' = \frac{1}{\rho^2\pi}\int_0^{X'} e^{-X'^2}dX' \,.\int_0^{Y'} e^{-Y'^2}dY'$$
$$- \frac{1}{3\rho^2\pi}\Big[K_{30}\int_0^{Y'} e^{-Y'^2}dY' \,.\{1-(1-2X'^2)e^{-X'^2}\}$$
$$+ 3K_{21}e^{-X'^2}.X'(1-e^{-Y'^2}) + 3K_{12}e^{-Y'^2}.\,Y'(1-e^{-X'^2})$$
$$+ K_{03}\int_0^{X'} e^{-X'^2}dX' \,.\{1-(1-2Y'^2)e^{-Y'^2}\}\Big].$$

In quoting this formula above on p. 37, the ′s are unnecessary and X, Y are written for $X'Y'$, and M_{32} &c. for K_{30} &c.

CHAPTER V

THE APPLICATION OF THE GENERALISED LAW OF ERROR

How far the law of frequency thus given can be obtained by experiment or is realised in nature can only be determined by examples. Experiments have been made and the results tested in cases where the genesis of the numbers is exactly known (throws of dice, draws of cards, etc.) and where it can be presumed, *e.g.* on the hypothesis that digits occur with equal frequency, and especially in the case of averages obtained by sample from a known frequency curve. Probably statistics such as birth-rates come in this class. In this group, where the constants (moments, etc.) are known exactly or approximately, the questions are whether the agreement is close having regard to the convergency of the successive approximations, which depends on m, the number of items going to each total and the values of j, i, etc.; and whether the resultant errors are within the limits to be anticipated from sampling, which depends on n, the number of separate experiments. A sub-case is when there is some correlation between the elements, as in hands of cards, and here the question is whether the amount of interdependence is sufficient to strain the general law of distribution.

There is no *a priori* reason to lead us to do more than assume as an hypothesis that frequency groups furnished by anthropometry or biology are generated in the manner postulated for the law of error. Perhaps in the case of shots at a target, or instrumental measurements of a physical quantity, the hypothesis of the addition of the effects of many independent causes is more plausible. The earlier astronomers' work in the latter field, and Quetelet, Galton and others in the former, had given examples where the normal curve was at least approximately attained. Edgeworth gives many cases where similar phenomena are examined in the light of the further approximations. Where the agreement is satisfactory, we can only say that the results are consistent with the hypothesis, and not that in fact the entities are built up in the manner postulated. Experimental results are to be found especially in (44), (45), (57) and (58). Many cases were found in which the agreement was not sufficient, and the method of translation, described below (p. 65) was devised in an endeavour to find a law, akin to the general law, which would include these. None of the tests extend beyond the third approximation.

We have to consider how the equation of the curve should be fitted to observations, for which purpose the two methods of moments and of percentiles are used; how the differences between the observations and the curve should be measured, whether by aggregate misfit or by Prof. Pearson's criterion; and what is the probability that so great an aggregate of errors should arise in sampling. There is also the method of so choosing the constants of the curve as to minimise the improbability of the emergence of the aggregate of the errors.

If we fit by moments, we at once determine the average, c, j and i by equating them to the corresponding quantities found from the observations. The integral form on p. 46 above is then at once available for making comparisons with the observations in as much detail as they allow. (45), pp. 335 *seq.* and (58), *J.S.S.*, 1917, p. 75.

There are three reasons why this method is not necessarily the best. First, the probable errors of j and i, as determined from n observations, involve respectively the sixth and eighth moments of the curve divided by $\sqrt{n}$, and consequently are subject to considerable errors of sampling, which may be of as great order as the terms neglected in the formula, which depend among other things on $m^{-\frac{3}{2}}$, where m is the (hypothetical) number of elements that go to form a single observation. Secondly, though in the normal curve the values of the average and of the modulus determined from the observations are those which lead to minimum improbability (see pp. 24–5 above), there is no such sanction for j and i. Thirdly, the observations may be in such wide or irregular grades that even with Dr. Sheppard's corrections the moments cannot be accurately determined, while the method of percentiles is particularly adapted to use the observations accurately however they are graded.

In applying criteria of fit it is to be remembered that as $\frac{x}{c}$ increases the coefficients in the neglected terms increase, to an extent that can be decided for given values of j and i, and that the distribution at the extremities of the curves is not theoretically known. When j is as great as ·1 the neglect of j^2 is only safe up to $x = \pm c$ approximately, and if i is as great as ·1 the neglect of i similarly tends to vitiate the comparison after the same margin. From $-c$ to $+c$ are included about 86% of the observations ((58), *J.S.S.*, 1917, p. 67). For values of j and i greater than ·1 Edgeworth in his last treatment preferred the method of translation (p. 65 below).

The *method of percentiles* depends upon the following simple formulæ :—

Transfer the equation to the second approximation to the median $(-\frac{1}{3}jc)$ as origin by replacing x by $x - \frac{1}{3}jc$.

$$y = \frac{1}{c\sqrt{\pi}} e^{-(x-\frac{1}{3}jc)^2/c^2} \{1 - 2j((x - \tfrac{1}{3}jc)/c - \tfrac{2}{3}(x - \tfrac{1}{3}jc)^3/c^3)\}$$

$$= (\text{after expansion and neglect of } j^2), \frac{1}{c\sqrt{\pi}} e^{-\frac{x^2}{c^2}} \left\{1 - \frac{4}{3}j\left(\frac{x}{c} - \frac{x^3}{c^3}\right)\right\}$$

$$\therefore \int_0^x y\,dx = \frac{1}{c\sqrt{\pi}} \int_0^x e^{-\frac{x^2}{c^2}} dx - \frac{2}{3}j\frac{x^2}{c^2}\cdot\frac{e^{-\frac{x^2}{c^2}}}{c\sqrt{\pi}} = \int_0^\xi \frac{1}{\sqrt{\pi}} e^{-\xi^2} d\xi, \text{ say.}$$

Now the left-hand side is the proportion of all the observations from the median to x, that is $\frac{1}{2}$ less the known proportion above x; hence ξ can be found from the tables of the normal function without knowing the median.

Write $\xi = (x + d)/c$. Then it is clear that d is of the order j, and d^2 is negligible. Then

$$\int_0^\xi \frac{1}{\sqrt{\pi}} e^{-\xi^2} d\xi = \int_0^{\frac{x}{c}} \frac{1}{c\sqrt{\pi}} e^{-\frac{x^2}{c^2}} dx + d\frac{1}{c\sqrt{\pi}} e^{-\frac{x^2}{c^2}}$$

$$\therefore d = -\frac{2}{3}j\frac{x^2}{c^2}, \text{ and } \xi = \frac{x}{c} - \frac{2}{3}j\frac{x^2}{c^2}.$$

Again neglecting j^2, $x = c\xi + \frac{2}{3}jc\xi^2$.

x itself is not known until the median is known, but the difference between say x_t and x_{t+1} is known, so that we have equations such as $x_{t+1} - x_t = c(\xi_{t+1} - \xi_t) + \frac{2}{3}jc({\xi_{t+1}}^2 - {\xi_t}^2)$, to determine c and j, and then the median can be found from any one observation. The three quantities needed can in fact be determined from three accurate observations only.

If the median and quartiles are adequately known by interpolation or otherwise, as the points M, Q_1, Q_2 on the axis of abscissa we have $MQ_2 = c\xi + \frac{2}{3}jc\xi^2$

$$MQ_1 = -\,c\xi + \tfrac{2}{3}jc\xi^2$$

where $\xi = \cdot 477\ldots$, the multiple that the probable error is of the modulus in the normal curve. Then $c = \dfrac{Q_1M + MQ_2}{2\xi}$, and

$$j = \frac{MQ_2 - Q_1M}{MQ_2 + Q_1M} \times \frac{3}{2\xi} = \text{approximately } \frac{MQ_2 - Q_1M}{MQ_2 + Q_1M} \times 3{\cdot}15.$$

The various methods just described are illustrated ((45), pp. 336 *seq.*) as follows from wage statistics.

Fitting by Moments

From the data the average is $\bar{x} = 24{\cdot}65$ shillings, $c = \sigma\sqrt{2} = 8{\cdot}70s.$, $\mu_3 = 157{\cdot}2$, $j = \mu_3 \div c^3 = \cdot 237$.

1. Upper limits of groups x.	2. $\tau = \frac{x-\bar{x}}{c}$	3. $\int_0^\tau \frac{1}{\sqrt{\pi}} e^{-\tau^2} . d\tau$	4. $\frac{1}{3\sqrt{\pi}}(1-2\tau^2)e^{-\tau^2}$	5. Col. 4 × j.	6. Sum of cols. 3 and 5.	7. Contents of grades. Differences in col. 6.	8. Contents of grades. Observations.	9. (Col. 7 ~ col. 8) × 1000.
10*s*.	−1·684	−·491*	−·051	−·012	−·503	—	·025	—
15*s*.	−1·109	−·440*	−·080	−·019	−·459	·041†	·025	16
20*s*.	−0·535	−·275*	·061	·014	−·261	·198	·215	17
25*s*.	0·040	·023	·187	·044	·067	·328	·336	8
30*s*.	0·615	·308	·031	·007	·315	·248	·242	6
35*s*.	1·189	·454	−·084	−·020	·434	·119	·116	3
40*s*.	1·764	·494	−·043	−·010	·484	·050	·042	8
				Maximum	·500	·016	·024	8
								66

* $\int_\tau^0$ below the average.

† Difference from −·500; for this irregularity see p. 76.

The "misfit" is 6·6%.

ALTERNATIVE COMPUTATION OF j

10. Grades.	11. Contents of grades × 1000. Calculated.	12. Observed.	13. Differences.	14. Order of magnitude.	15. Col. 13 when $j = \cdot 302$.
Up to 15*s*.	$60 - 80j$	25	$80\,(j - \cdot 43)$*	i	11
15*s*.—20*s*.	$165 - 141j$	215	$141\,(j - \cdot 35)$	iii	7·4
20*s*.—25*s*.	$298 - 126j$	336	$126\,(j - \cdot 30)$	iv	0
25*s*.—30*s*.	$285 - 156j$	242	$156\,(j - \cdot 28)$	v	3·4
30*s*.—35*s*.	$146 - 115j$	116	$115\,(j - \cdot 26)$	vi	4·7
35*s*.—40*s*.	$40 - 41j$	42	$41\,(j - \cdot 00)$	vii	10·4
40*s*. and above	$6 - 43j$	24	$43\,(j - \cdot 41)$	ii	5
					42

Col. 11 is what col. 7 (× 1000) becomes when j is not substituted for, and the extreme limits are taken as ± 500.

We proceed by the method of the weighted median, which is further discussed pp. 103 *seq.* below. We have seven values of j to choose from. The aggregate of the difference (and therefore the percentage misfit) is least when the value is chosen in the grade which contains the (weighted) median difference. The sum of the numbers outside the brackets in column 13 is 702; of these i, ii and iii contain 264 (less than half), and i, ii, iii and iv contain 390 (more than half). The least misfit is found by taking the value in grade iv, viz. $j = \cdot 30$ (more exactly $\cdot 302$); for any positive divergence from this, say ϵ, adds 438ϵ to the positive errors in grades iv to vii and substracts only 264ϵ from i, ii and iii; and similarly if ϵ is negative the aggregate is increased.

The misfit is now reduced to 4·2%.

COMPUTATION BY PERCENTILES

Limits.	Observations.	Proportion between median and	$= F(\xi)$.	Hence $\xi =$	x.
Less than 10*s*.	0·001	10*s*.	·499	−2·200	0
10*s*.—15*s*.	0·024	15*s*.	·475	−1·385	1
15*s*.—20*s*.	0·215	20*s*.	·260	−0·499	2
20*s*.—25*s*.	0·336	25*s*.	·076	0·136	3
25*s*.—30*s*.	0·242	30*s*.	·318	0·642	4
30*s*.—35*s*.	0·196	35*s*.	·434	1·065	5
35*s*.—40*s*.	0·042	40*s*.	·476	1·400	6
40*s*. and above	0·024				
	1·000				

* $25 - (60 - 80j) = 80j - 35 = 80\,(j - \cdot 43)$.

Take 5*s*. as unit, 10*s*. as origin, so that x has the values in the final column. The equation

$$x_{t+1} - x_t = c(\xi_{t+1} - \xi_t) + \tfrac{2}{3}jc(\xi_{t+1}{}^2 - \xi_t{}^2)$$

becomes, since $x_{t+1} - x_t = 1$,

$$c + \tfrac{2}{3}jc(\xi_{t+1} + \xi_t) = \frac{1}{\xi_{t+1} - \xi_t}.$$

Write $c = u$, $\tfrac{2}{3}jc = v$, and we have (working from 40*s*. down):

$$u + 2{\cdot}47v = \frac{1}{{\cdot}335} = 2{\cdot}99 \ (a)$$

$$u + 1{\cdot}71v = \frac{1}{{\cdot}423} = 2{\cdot}36 \ (b)$$

$$u + 0{\cdot}78v = \frac{1}{{\cdot}514} = 1{\cdot}97 \ (c)$$

$$u - 0{\cdot}36v = \frac{1}{{\cdot}635} = 1{\cdot}57 \ (d)$$

$$u - 1{\cdot}88v = \frac{1}{{\cdot}886} = 1{\cdot}13 \ (e)$$

$$u - 3{\cdot}58v = \frac{1}{{\cdot}815} = 1{\cdot}22 \ (f).$$

The lines representing these equations in u and v would be concurrent if the fit were perfect.

Edgeworth's double median method (discussed pp. 103 *seq.* below) is employed to give an approximation, and it leads to the intersection of lines (*b*) and (*e*), giving $c = u = 1{\cdot}77$, *i.e.* 8·85 shillings; $v = {\cdot}343 = \tfrac{2}{3}jc$; $j = {\cdot}29$.

The values of X given by the formula $c\xi + \tfrac{2}{3}jc\xi^2$ are

				Hence median.	Add $\tfrac{1}{3}jc$ = ·885 shillings.
At 10*s*.	−2·23	× 5 =	−11·15*s*.	21·15	22·0
15*s*.	−1·79	,,	− 8·95*s*.	24·95	25·8
20*s*.	−0·130	,,	− 0·65*s*.	20·65	21·5
25*s*.	0·247	,,	1·12*s*.	23·88	24·8
30*s*.	1·277	,,	6·38*s*.	23·62	24·5
35*s*.	2·274	,,	11·37*s*.	23·63	24·5
40*s*.	3·250	,,	16·25*s*.	23·75	24·6

24·65*s*., the known centre of gravity of the observations, is as good a value as we can get from these. Working through the first table (p. 58 above) with this average and the new values of j and c we obtain a misfit of about 5%.

These methods of a somewhat arbitrary character for determining the constants may be stumbling-blocks to statisticians who are used to the clean-cut methods of moments and of least squares. They are, however, in the case before us unnecessary if moments can be adequately determined; if they cannot, or if their use is unjustified, then in all cases supplementary methods must be devised.

Method of minimising Improbability of Errors

A less arbitrary method is to extend the principle, already named (pp. 16 *seq.*), used for justifying the computation of the average and modulus in the normal law. Let us seek the values for all the constants by the condition that they should make the improbability of the observations arising at random from a curve of the given form a minimum. Now Prof. Pearson has shown that the measure of this improbability depends on $\sum \frac{e^2}{pN}$, where N is the number of observations, p the proportion in the frequency curve from which they are obtained in a random sample in a given grade or at a given point, and e the difference between pN and $p'N$, the number of observations, the sum being taken over all the grades or points. When this expression is a minimum, so is the improbability. We may without loss of generality omit the N and treat the e's as any (the same) proportion of the actual errors. Edgeworth, however, uses $\sum \frac{e^2}{p'}$ instead of $\sum \frac{e^2}{p}$, since p' is known but not p, and $p' \sim p$ is of an order that may be neglected in the working of his analysis.

The constants needed are $\bar{x}$, c, j and i. $\bar{x}$ and c are finite, j is of order $\frac{1}{\sqrt{m}}$, i of order $\frac{1}{m}$. Approximate values of $\bar{x}$ or of the median and of c $\left(\text{or rather of } h = \frac{1}{c}\right)$ are first obtained so that the corrections to them are of the same order as j ((58), *J.S.S.*, 1916, pp. 486 and 497).

If we take as a simplified example the case where the median and modulus are adequately determined from the observations, the method of finding j and i is as follows. Write the integral equation for the generalised law in the form

$$Y_\tau = \int_0^\tau z d\tau = F(\tau) + j\phi(\tau) + i\psi(\tau),*$$

where τ is $\dfrac{x - \overline{m}}{c}$, x being the abscissa and $\overline{m}$ the median as it appears in the data,

$$F(\tau) = \int_0^\tau \frac{1}{\sqrt{\pi}} e^{-\tau^2}, \quad \phi(\tau) = \frac{2\tau^2}{3} \cdot \frac{e^{-\tau^2}}{\sqrt{\pi}},$$

and $\psi(\tau) = (-\tfrac{1}{2}\tau + \tfrac{1}{3}\tau^3)\dfrac{e^{-\tau^2}}{\sqrt{\pi}}$.

Let U_r be the proportion observed between values τ_1 and τ_2 corresponding to the x's that determined the rth grade in the data, and E_r the error due to sampling.

Then

$$\begin{aligned} E_r &= F(\tau_2) - F(\tau_1) + j\{\phi(\tau_2) - \phi(\tau_1)\} + i\{\psi(\tau_2) - \psi(\tau_1)\} - U_r \\ &= \text{(say) } K_r + jL_r + iM_r - U_r. \end{aligned}$$

We have then to determine the values of j and i which minimise $\sum \dfrac{E_r^2}{U_r}$, the sum extending either over all the grades, or omitting the extreme grades where the law is not expected to prevail ((58), p. 487). All the quantities other than j and i are known.

We have

$$\frac{\partial}{\partial j}\left(\sum \frac{E_r^2}{U_r}\right) = 0 \quad \therefore j\sum \frac{L_r^2}{U_r} + i\sum \frac{L_r M_r}{U_r} = \sum \frac{L_r(U_r - K_r)}{U_r},$$

and

$$\frac{\partial}{\partial i}\left(\sum \frac{E_r^2}{U_r}\right) = 0 \quad \therefore j\sum \frac{L_r M_r}{U_r} + i\sum \frac{M_r^2}{U_r} = \sum \frac{M_r(U_r - K_r)}{U_r}.$$

The work of finding the quantities U_r, K_r, L_r and M_r is shown for seven compartments ((58), *J.S.S.*, 1916, p. 490), and is not at all prohibitive.

Edgeworth, in fact, proceeds by first determining j when i is

* Edgeworth frequently omits the j^2 term, apparently on the ground that in several examples i is little less than j, and j^2 negligible as well as i^2. He gives his reasons more fully in his paper ((71), pp. 578–9).

ignored, and then determining i with the value of j so given as constant. The results are the same only if $\left(\sum \frac{LM}{U}\right)^2$ is negligible compared with $\sum \frac{L^2}{U} \times \sum \frac{M^2}{U}$, and it may be suggested that the very slight reduction in the numerical work is not worth saving. With adequate tables the work so far requires only a little patience.

With these values for j and i we can proceed by methods which involve no new principles to adjust the values of $\bar{m}$ and c; and then again to compute j and i; but this addition to the work adds materially to the time needed, and in fact $\bar{m}$ and c can often be determined with sufficient accuracy by simple processes, or if need be by methods given in the same paper ((58), pp. 486, 492).

Limits of Applicability

The question, over what range the law of error (including the j and i terms) gives a sufficient approximation, naturally does not admit of an exact answer. Dealing first with j ((58), *J.S.S.*, 1917, pp. 66 *seq*), write $\chi(j)$ for $e^{-\frac{1}{6}jD^3}\left(\frac{1}{\sqrt{\pi}}e^{-\tau^2}\right)$.

$$\chi(j) = \chi(0) + j\chi'(0) + \tfrac{1}{2}j^2\chi''(j_1), \text{ where } \chi'(0) = \frac{d}{dj}\cdot\chi(j),\ j = 0,$$

for the first two terms give the second approximation with unit modulus and the third is the remainder in Taylor's expansion, j_1 being between 0 and j.

$$\frac{1}{2}\chi^2(j_1) = \frac{1}{72}D^6 \,.\, e^{-\frac{1}{6}j_1D^3}\left(\frac{1}{\sqrt{\pi}}e^{-\tau^2}\right) = \frac{1}{72}D^6\left(\frac{1}{\sqrt{\pi}}e^{-\tau^2}\right) + a$$

where a is presumably a small quantity. Now

$$\frac{1}{72}D^6\left(\frac{1}{\sqrt{\pi}}e^{-\tau^2}\right) = \frac{1}{72}(-120 + 720\tau^2 - 480\tau^4 + 64\tau^6)\,.\,\frac{1}{\sqrt{\pi}}e^{-\tau^2}$$

approximately.

The *relative* addition to the ordinate of the normal curve due to the neglect of j^2 is therefore approximately

$$j^2 \times \tfrac{1}{9}(-15 + 90\tau^2 - 60\tau^4 + 8\tau^6).$$

Some values of this expression are

τ	j ·2	j ·1	j ·05
1	·100	·025	·0062
2	— ·46	— ·114	— ·0285

Hence when j is as great as ·1 we may proceed as far as $\tau = \pm i$, that is as far as the modulus on either side of the average, taking in about 86% of the observations; if j is no greater than ·05 we may proceed to twice the modulus and so include practically all the observations.

Proceeding similarly for i, taking j as known, the relative correction is approximately for selected values

τ	i ·2	i ·1	i ·05
1	·06	·015	·004
2	·032	·008	·002

The neglect of i^2 is therefore somewhat safer than the neglect of j^2, for equal values of j and i.

On the question of over what range the generalised law may be expected to give a good fit, see also pp. 76–7 below.

CHAPTER VI

THE METHOD OF TRANSLATION

THOUGH the trials made show that the second approximation represents adequately, up to and beyond the limits just discussed, a great number of frequency groups of different kinds, it is found that its sphere by no means extends to all groups where the main postulate of many independent causes might be expected to hold.

Edgeworth divides groups into three classes, normal, subnormal and abnormal: normal where j and i are negligible, subnormal where j and i are significant but do not exceed about ·1, abnormal where the values are greater. The division is of course not rigid.

Abnormal groups may have so great asymmetry as to be nearly unilateral, the mode corresponding to an impossible negative abscissa, or have unusual kurtosis so as to be even U-shaped. Such groups have no evident affinity to the normal curve, but in fact some of the modifications suggested in its form are capable of being applied to them.

Subnormal groups may either arise from elements as postulated for the general law of error, but have so great values of j and i that the approximation breaks down not very far from the centre; or the postulate of independence or of the aggregate being formed by a function allied to simple summation may not be valid, even when extended as on pp. 30, 34 and 40 above. To deal with such cases it may be investigated whether revised postulates lead to any general expression which is found on trial to account for observational groups, and the "method of translation" is developed by Edgeworth for this purpose.

In the earlier papers he was inclined to contrast his method of translation with Prof. Pearson's well-known system, claiming that his own was based on a presumed *vera causa*, while the latter was purely empirical. More recently he was inclined to make that claim for the generalised law of error only, and to admit that translation had the same kind of validity as had Prof. Pearson's hypothesis, and to depend rather on the criteria of convenience of working and success in application (see pp. 81 *seq.* below).

The idea of translation was introduced in 1898 somewhat as follows ((44), p. 675). It is known that the frequency group of men's heights (in a homogeneous class) is normal. If weight was proportional to the cube of height, the distribution of weights would not be exactly normal. If the curve for heights is $\frac{1}{c\sqrt{\pi}}e^{-\frac{(x-g)^2}{c^2}}$, and the relation between weight and height is $w = kx^3$, then for any height g (the average) $+ d$, the weight is

$$w = k(g+d)^3 = kg^3\left\{1 + \frac{3d}{g}\left(1 + \frac{d}{g} + \frac{d^2}{3g^2}\right)\right\}.$$

Now the modulus for height is about 3 inches, and $\frac{c}{g} =$ about $\frac{1}{22}$. d is only a small multiple of c (very rarely exceeding 2).

If, for example, $d = c$, then with these numbers

$$w = kg^3 + 3kg^2\frac{d}{22}\left(1 + \frac{1}{22} + \frac{1}{1452}\right).$$

A first approximation to the frequency curve of w is obtained by neglecting $\frac{d^2}{g^2}$ in comparison with $\frac{d}{g}$. Then $w = kg^3\left(1 + \frac{3d}{g}\right)$. The curve for d is $\frac{1}{c\sqrt{\pi}}e^{-\frac{d^2}{c^2}}$. Therefore the curve for w is normal, with deviation $\delta = 3d \times \frac{\text{Average weight}}{\text{Average height}}$.* The modulus is $3kg^2 \times c$.

To find the second and third approximations we " translate." Refer the height distribution to its average, writing ξ for $d = x - g$, and further write $x = w - kg^3$; we have

$$x = 3kg^2\left(\xi + \frac{\xi^2}{g} + \frac{\xi^3}{3g^2}\right).$$

* Neglecting the term of order $\frac{d^2}{g^2}$.

Now take unit modulus for the curve of heights ($c = 1$), then $3kg^2 = a$, say, is the modulus for the first approximation for x, and $x = a\left(\xi + \frac{\xi^2}{g} + \frac{\xi^3}{3g^2}\right)$.

This is generalised to $x = f(\xi)$, which if $f(\xi)$ is rapidly convergent may be written with appropriate origin $x = a(\xi + \kappa\xi^2 + \lambda\xi^3)$, where κ and λ are small, and a finite, and the frequency distribution of ξ is $\eta = \frac{1}{\sqrt{\pi}}e^{-\xi^2}$ (called the generating curve). It is suggested that the frequency of x thus generated may be common in practical statistics. The ordinate η corresponding to any ξ is to be shifted (or translated) to a corresponding abscissa x, and the new curve so obtained is the translated curve.

The scale of abscissa is modified so that a small area $\eta\Delta\xi$ is translated as an area $y\Delta x$, where $\Delta x = a(1 + 2\kappa\xi + 3\lambda\xi^2)\Delta\xi$. The new curve is of course unsymmetrical unless $\kappa = 0$. The mean of the ξ-curve becomes the median of the new, and each percentile of the ξ-curve becomes the same percentile of the new.

Thus take the equation

$$\begin{aligned} x &= 2(\xi + \cdot 2\xi^2 + \cdot 1\xi^3) \\ \Delta x &= 2(1 + \cdot 4\xi + \cdot 3\xi^2)\Delta\xi \\ y\Delta x &= \eta\Delta\xi. \end{aligned}$$

The values at the deciles of the normal curve $\eta = \frac{1}{\sqrt{\pi}}e^{-\xi^2}$ are given in the first columns and the corresponding values of x, y in the latter columns.

	ξ	η	x	y
D_9	·905	·249	2·28	·077
D_8	·595	·396	1·34	·148
D_7	·370	·492	0·80	·206
D_6	·179	·546	0·37	·250
M	—	·564	—	·283
D_4	− ·179	·546	− 0·35	·292
D_3	− ·370	·492	− 0·70	·280
D_2	− ·595	·396	− 1·03	·228
D_1	− ·905	·249	− 1·63	·139

The average of the translated curve is shown on p. 69 below to be at $+\frac{1}{2}a\kappa = \cdot 2$. The medians of the curves coincide.

The diagram shows the companion curves.*

. . . ABC . . . represents the curve $\eta = \frac{1}{\sqrt{\pi}} e^{-\xi^2}$.

. . . abc . . . represents the translated curve where

$$x = 2(\xi + \cdot 2\xi^2 + \cdot 1\xi^3).$$

The vertical lines through D_1 . . . M . . . D_9 and through d_1 . . . M . . . d_9 divide the curves respectively into 10 equal areas.

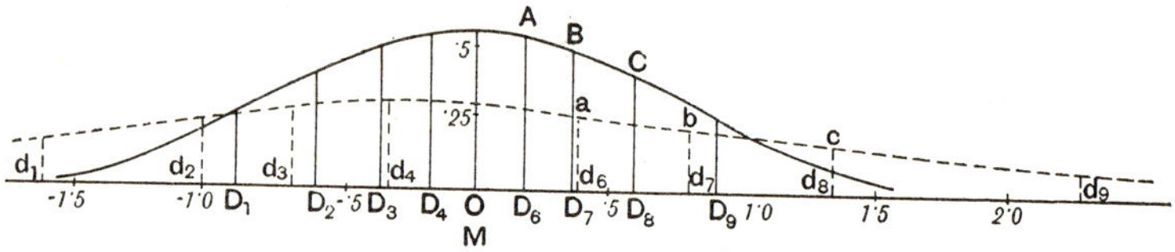

Such an area as D_6ABD_7 is translated to d_6abd_7.

We have to determine from the observations appropriate values of the median, to be taken as origin, of a which is related to the modulus, and of κ and λ. This may be done by the method of moments, or the method of percentiles, or by consideration of what constants give the best fit, as in the case of the generalised law of error. But here the method of percentiles has a special appropriateness from the very nature of the assumed generation of the statistics.

Method of Moments (56) and (57)

The tth moment (M_t) of the x-curve about its median is obtained by multiplying each element of area $\eta d\xi$ by x^t, adding or integrating, and dividing by the area.

$$M_t = \int_{-\infty}^{\infty} x^t \,.\, \eta d\xi = \int_{-\infty}^{\infty} a^t(\xi + \kappa\xi^2 + \lambda\xi^3)^t \frac{e^{-\xi^2}}{\sqrt{\pi}} d\xi.$$

An even moment of the generating curve,

$$\int_{-\infty}^{\infty} \xi^{2l} \frac{e^{-\xi^2}}{\sqrt{\pi}} d\xi = \frac{(2l)!}{2^l \,.\, l!} \cdot \left(\frac{1}{2}\right)^l,$$

since its standard deviation is $\frac{1}{\sqrt{2}}$. Give t values 0 to 4 :

* Edgeworth gives a diagram, but not the detailed working.

$$M_0 = 1;\ M_1 = \tfrac{1}{2}a\kappa;\ M_2 = a^2(\tfrac{1}{2} + \tfrac{3}{4}\kappa^2 + \tfrac{3}{2}\lambda + \tfrac{15}{8}\lambda^2);$$
$$M_3 = a^3(\tfrac{9}{4}\kappa + \tfrac{15}{8}\kappa^3 + \tfrac{45}{4}\kappa\lambda + \tfrac{315}{16}\kappa\lambda^2);$$
$$M_4 = a^4(\tfrac{3}{4} + \tfrac{45}{4}\kappa^2 + \tfrac{105}{16}\kappa^4 + \tfrac{15}{2}\lambda + \tfrac{315}{4}\kappa^2\lambda + \tfrac{315}{8}\lambda^2 + \tfrac{10395}{64}\lambda^4 + \tfrac{2835}{16}\kappa^2\lambda^2 + \tfrac{945}{8}\lambda^3).$$

The mean is therefore $\frac{1}{2}a\kappa$ to the right of the median.

Transfer these moments by the usual formulæ to give them about the mean instead of about the median, and equate them to the moments μ_2, μ_3, μ_4 computed from the observations.

$$\mu_2 = M_2 - M_1^2 = \frac{a^2}{2}\left(1 + \kappa^2 + 3\lambda + \frac{15}{4}\lambda^2\right);$$
$$\mu_3 = a^2\kappa\left(\frac{3}{2} + \kappa^2 + 9\lambda + \frac{135}{8}\lambda^2\right);$$
$$\mu_4 - 3\mu_2^2 = a^4\left(6\kappa^2 + 3\lambda + 3\kappa^4 + 54\kappa^2\lambda + 27\lambda^2 + 135\kappa^2\lambda^2 + \frac{405}{4}\lambda^3 + \frac{1215}{8}\lambda^4\right).$$

We have three equations to determine a, κ and λ.

Since no odd power of κ occurs, write $\kappa^2 = \chi$.

Further write $\frac{1}{8}\cdot\frac{\mu_3^2}{\mu_2^3} = j^2$ and $i = \frac{\mu_4}{4\mu_2^2} - \frac{3}{4}$, as in the former notation (p. 31). Eliminate a from the equations, and we have

$$\chi(\tfrac{3}{2} + \chi + 9\lambda + \tfrac{135}{8}\lambda^2)^2 - j^2(1 + \chi + 3\lambda + \tfrac{15}{4}\lambda^2)^3 = 0;$$
$$6\chi + 3\lambda + 3\chi^2 + 54\chi\lambda + 27\lambda^2 + 135\chi\lambda^2 + \tfrac{405}{4}\lambda^3 + \tfrac{1215}{8}\lambda^4 - i(1 + \chi + 3\lambda + \tfrac{15}{4}\lambda^2)^2 = 0.$$

When these are solved, we compute a^2 from the earlier formula, and the median is $\frac{1}{2}a\kappa$ to the left of the observed mean.

A table giving the values of $\beta = 8j^2$ and $\epsilon = \frac{1}{3}i$ * from a range of values of χ 0 to ·06 and of λ + ·10 to − ·06 was given in the *Statistical Journal*, 1924, p. 582 (71). This made it possible to locate approximately values of χ and λ which correspond to given j^2 and i, and thence by interpolation or use of the equations above to obtain χ and λ fairly rapidly. A reversed table showing χ and λ for given β and ϵ has now been computed

* Edgeworth uses β' for j^2 and ϵ for $\frac{i}{3}$.

over a greater range and with more detail, and is produced at the end of the text, pp. 122–9 below. It is now a very short matter to compute the constants from the moments, and one of the principal practical objections to the method of translation would be removed.

If by the table named or otherwise it is ascertained that χ and λ are small, so that χ^2, $\chi\lambda$ and λ^2 are negligible, we have at once $\chi = \frac{4}{9}j^2$; $\kappa = \frac{2}{3}j$; and $\lambda = \frac{i}{3} - \frac{8}{9}j^2$. . . . These values make a first, and often a sufficient, approximation for χ and λ. A second approximation, when χ^3, $\chi^2\lambda$, $\chi\lambda^2$ and λ^3 are neglected, is

$$\chi = \frac{4}{9}j^2 + \frac{368}{243}j^4 - \frac{4}{9}ij^2; \quad \lambda = \frac{i}{3} - \frac{8}{9}j^2 - \frac{784}{243}j^4 + \frac{56}{27}ij^2 - \frac{1}{3}i^2,$$

but it is found that these values do not give good results unless j^2 and i are quite small.

The constants being determined, then the equation in ξ

$$a(\xi + \kappa\xi^2 + \lambda\xi^3) = x$$

must be solved for each value of x, when x is the abscissa of the boundary of a grade (or of a discrete reading) given in the observations, with the median as origin.

Approximate solutions are given by

$$\xi = \frac{x}{a}, \quad -\frac{x^2}{a^2}\left(\kappa + \lambda\frac{x}{a}\right), \quad +\frac{x^3}{a^3}\left(\kappa + \lambda\frac{x}{a}\right)\left(2\kappa + 3\lambda\frac{x}{a}\right).$$

Then $\int_{\xi_1}^{\xi_2} \frac{1}{\sqrt{\pi}} e^{-\xi^2} d\xi$ gives the proportion of the observations to be expected between x_1 and x_2, the values of x corresponding to ξ_1, ξ_2.

Thus in an example given ((71), p. 583) the arithmetic mean is 367·62, $\sqrt{\mu_2} = 39{\cdot}98$, $j^2 = {\cdot}0375$, $i = {\cdot}3375$. Hence $\chi = {\cdot}01455$, $\kappa = {\cdot}1206$, $\lambda = {\cdot}0682$, $a = 50{\cdot}85$. N, the number of observations, $= 200$. The constructive median is $367{\cdot}62 - \frac{1}{2}\kappa a = 364{\cdot}554$.

$$\xi = \frac{x}{a} - \left(\frac{x}{a}\right)^2\left({\cdot}1206 + {\cdot}068\frac{x}{a}\right) + \ldots,$$

or the equation $\cdot0682\xi^3 + \cdot1206\xi^2 + \xi = \frac{x}{a}$ can be solved.

FREQUENCY OF BACILLI

Limits of grades (No. of bacilli).	Distance from median. x	$\frac{x}{a}$	ξ	$\int_0^\xi \frac{1}{\sqrt{\pi}} e^{-\xi^2} . d\xi$.	Differences $\times$ 200. ν	Error. e	$\frac{e^2}{\nu}$
330	$-34{\cdot}054$	$-{\cdot}670$	$-{\cdot}706$	$-$ ·341*	$100 - 68{\cdot}2 = 31{\cdot}8$	3·8	0·46
346	$-18{\cdot}054$	$-{\cdot}355$	$-{\cdot}368$	$-$ ·199*	$68{\cdot}2 - 39{\cdot}7 = 28{\cdot}5$	0·5	0·01
362	$-2{\cdot}054$	$-{\cdot}040$	$-{\cdot}041$	$-$ ·023*	$-4{\cdot}6 + 39{\cdot}7 = 35{\cdot}1$	0·9	0·02
378	13·946	·274	·265	·146	$29{\cdot}2 + 4{\cdot}6 = 33{\cdot}8$	0·2	0·00
394	29·946	·589	·542	·279	$55{\cdot}7 - 29{\cdot}2 = 26{\cdot}5$	7·5	2·10
					$100 - 55{\cdot}7 = 44{\cdot}3$	5·3	0·63
							3·23

Hence P, the probability of so great an aggregate of errors, deduced from *Tables for Statisticians and Biometricians*, p. 26, $n' = 6$, is ·65.

$$* \int_\xi^0 \frac{1}{\sqrt{\pi}} e^{-\xi^2} d\xi.$$

If we consider those sub-normal curves in which i^2 and j^2, and therefore κ^2, λ^2 and $\kappa\lambda$ are negligible, it is found that the equation resulting from translation is identical with the generalised law when the same quantities are neglected † ((58), J.S.S., 1917, p. 72). For the translated equation is $y\Delta x = \frac{1}{\sqrt{\pi}} e^{-\xi^2} \Delta\xi$, where

$$x = a(\xi + \kappa\xi^2 + \lambda\xi^3) \text{ and } a = c \div (1 + \kappa^2 + 3\lambda + \tfrac{15}{4}\lambda^2)^{\frac{1}{2}}.$$

We have to the order of approximation named,

$$\frac{1}{a} = \frac{1}{c}\left(1 + \frac{3\lambda}{2}\right)$$

$$\xi = \frac{x}{a} - \frac{x^2}{a^2}\left(\kappa + \lambda\frac{x}{a}\right) = \frac{x}{c}\left(1 + \frac{3\lambda}{2}\right) - \kappa\frac{x^2}{c^2} - \lambda\frac{x^3}{c^3}$$

$$\frac{d\xi}{dx} = 1 \div (1 + 2\kappa\xi + 3\lambda\xi^2)a = (1 - 2\kappa\xi - 3\lambda\xi^2)\left(1 + \frac{3\lambda}{2}\right)\frac{1}{c}$$

$$= \left(1 + \frac{3\lambda}{2} - 2\kappa\frac{x}{c} - 3\lambda\frac{x^2}{c^2}\right) \div c$$

$$e^{-\xi^2} = e^{-\left\{\frac{x}{c} + \lambda\frac{x}{c}\left(\frac{3}{2} - \frac{x^2}{c^2}\right) - \kappa\frac{x^2}{c^2}\right\}^2} = e^{-\frac{x^2}{c^2}}\left\{1 - 2\lambda\left(\frac{3}{2} - \frac{x^2}{c^2}\right)\frac{x^2}{c^2} + 2\kappa\frac{x^3}{c^3}\right\}$$

† This differs from the third approximation in that the term in j^2 is neglected. It appears, by an unverified approximation, that if j^2 (and κ^2) are retained the two curves do not coincide.

$$\therefore\ y = \frac{1}{c\sqrt{\pi}} e^{-\frac{x^2}{c^2}} \Big\{1 - 2\lambda\Big(\frac{3}{2} - \frac{x^2}{c^2}\Big)\frac{x^2}{c^2} + 2\kappa\frac{x^3}{c^3}\Big\}$$
$$\Big(1 + \frac{3\lambda}{2} - 2\kappa\frac{x}{c} - 3\lambda\frac{x^2}{c^2}\Big)$$
$$= \frac{1}{c\sqrt{\pi}} e^{-\frac{x^2}{c^2}} \Big\{1 - 2\kappa\Big(\frac{x}{c} - \frac{x^3}{c^3}\Big) + \lambda\Big(\frac{3}{2} - 6\frac{x^2}{c^2} + 2\frac{x^4}{c^4}\Big)\Big\}.$$

Now write

$$\kappa = \frac{2}{3}j,\ \lambda = \frac{i}{3} - \frac{8}{9}j^2 = \frac{i}{3}.$$
$$y = \frac{1}{c\sqrt{\pi}} e^{-\frac{x^2}{c^2}} \Big\{1 - \frac{4}{3}j\Big(\frac{x}{c} - \frac{x^3}{c^3}\Big) + i\Big(\frac{1}{2} - \frac{2x^2}{c^2} + \frac{2}{3}\cdot\frac{x^4}{c^4}\Big)\Big\}.$$

This is referred to the median. Write $x + \frac{1}{3}jc$ for x, so as to refer to the arithmetic mean, and expand and collect terms.

$$y = \frac{1}{c\sqrt{\pi}} e^{-\frac{x^2}{c^2}} \Big\{1 - 2j\Big(\frac{x}{c} - \frac{2}{3}\cdot\frac{x^3}{c^3}\Big) + i\Big(\frac{1}{2} - \frac{2x^2}{c^2} + \frac{2}{3}\cdot\frac{x^4}{c^4}\Big)\Big\}$$

the equation of p. 31, excluding the j^2 term.

Since the method of translation is "practically equivalent to the generalised law" "throughout the whole tract for which it is valid," it is unnecessary to use the generalised law, even for the central portion, when we are examining the results of translation, and consequently the method of translation can generally be used in examining the observational frequency groups with which we are concerned.

The Method of Percentiles ((57), pp. 724 *seq.*)

If $\frac{1}{2} - p$ is the proportion of observations above a reading x measured from the median of the observations, then the corresponding ξ is at once given by $p = \int_0^\xi \frac{1}{\sqrt{\pi}} e^{-\xi^2} . d\xi$ and $x = a(\xi + \kappa\xi^2 + \lambda\xi^3)$.

The median is not known, but the difference $x_2 - x_1$ between any two positions on the scale of observations is known, and we have a number of equations such as

$$x_{t+1} - x_t = a(\xi_{t+1} - \xi_t)\{1 + \kappa(\xi_{t+1} + \xi_t) + \lambda(\xi_{t+1}{}^2 + \xi_{t+1}\xi_t + \xi_t{}^2)\}$$

from which the required constants a, κ and λ can be determined. If we can get sufficiently accurate values of the median and any three percentiles from the observations, a, κ and λ can be readily determined.

Determination of Constants by minimising Errors

When this method is reviewed it will be found that it is necessary first to determine the median. Since in the central portion of the curve the aberration from the normal curve is small (the coefficients of j and i being small), it is generally sufficient to determine $-\xi_1$ and ξ_2 which correspond to scale readings nearest to the median on the left and right (say $-x_1$ and x_2, where x_1+x_2 is known), and to write

$$\frac{x_1}{x_2}=\frac{\xi_1-\kappa\xi_1{}^2+\lambda\xi_1{}^3}{\xi_2+\kappa\xi_2{}^2+\lambda\xi_2{}^3}.$$

$\frac{x_1}{x_2}=\frac{\xi_1}{\xi_2}$ is a good approximation if the grades are narrow and therefore ξ_1 and ξ_2 small ((57), p. 732). If not, it is necessary to obtain approximate values of κ and λ by any of the methods already given.

The observations being regarded as a random sample from a curve whose form is known, but whose determining constants are not known, an x_r measured from the hypothetical median does not correspond exactly to the value, say $X_r=a(\xi_r+\kappa\xi_r{}^2+\lambda\xi_r{}^3)$ arising from the generating curve, where $\frac{1}{2}-\int_0^{\xi_r}\frac{1}{\sqrt{\pi}}e^{-\xi^2}.\,d\xi$ equals the proportion above ξ_r.

Write η_r for the ordinate of the generating curve, and therefore $\eta_r\left(\frac{d\xi}{dX}\right)_r=\eta_r a(1+2\kappa\xi_r+3\lambda\xi_r{}^2)$ for the ordinate of the translated curve.

Write x_r, x_{r+1} for the abscissæ of the limits of any grade in the observations, and let U_r' be the observed proportion in the grade. Let U_r be the constructed proportion in the grade X_r to X_{r+1}. Write $E_r=U_r'-U_r$ for the error in this area.

In the figure $Mn_r=x_r$, $Mn_{r+1}=x_{r+1}$, $MN_r=X_r$, $MN_{r+1}=X_{r+1}$, $n_rp=\eta_r$, $n_{r+1}q=\eta_{r+1}$, $U_r=N_rPQN_{r+1}$, $U_r'=n_rpqn_{r+1}$.

$$\text{Then } E_r = n_r pq n_{r+1} - N_r PQ N_{r+1}$$
$$= n_r N_r \,.\, N_r P - n_{r+1} N_{r+1} \,.\, N_{r+1} Q \text{ approx.}$$
$$= (X_r - x_r)\eta_r\left(\frac{d\xi}{dx}\right)_r - (X_{r+1} - x_{r+1})\eta_{r+1}\left(\frac{d\xi}{dx}\right)_{r+1}.$$

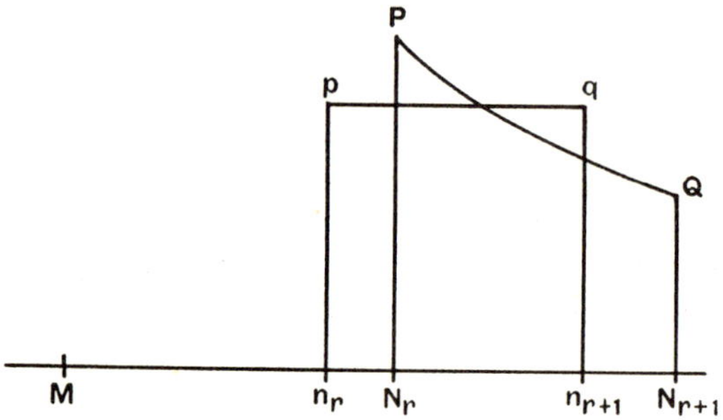

Now generally write $\Delta_r(f) = f_{r+1} - f_r$ for any function f, as in the notation of finite differences.

Then $E_r = \Delta_r\left(\dfrac{h\eta x - \eta\xi(1 + \kappa\xi + \lambda\xi^2)}{1 + 2\kappa\xi + 3\lambda\xi^2}\right)$, where $h = \dfrac{1}{a}$.

Here h is finite and κ and λ are small quantities. Neglect κ^3, λ^3, etc., expand the denominator and collect terms, and we obtain :

$$E_r = \Delta_r\{h\eta x(1 - 2\kappa\xi - 3\lambda\xi^2 + 4\kappa^2\xi^2 + 12\kappa\lambda\xi^2 + 9\lambda^2\xi^4)$$
$$- \eta\xi(1 - \kappa\xi - 2\lambda\xi^2 + 2\kappa^2\xi^2 + 7\kappa\lambda\xi^3 + 6\lambda^2\xi^4)\}.$$

Now, as on p. 61 above, the minimum of

$$\Sigma\frac{E_r^2}{U_r} = \Sigma\frac{E_r^2}{U_r' - E_r} = \Sigma\frac{E_r^2}{U_r''},$$

neglecting $\dfrac{E^3}{U}$, gives the unknown quantities which make the improbability of the sample arising from the supposed frequency group a minimum: the Σ is to be extended over all the compartments.

Determine a preliminary value of h, say h_0, when κ and λ are zero.

$$0 = \frac{d}{dh}\Sigma\left\{\frac{\Delta_r(h\eta x - \eta\xi)^2}{U_r}\right\} \text{ whence}$$

$$h_0\Sigma\frac{(\Delta_r\eta x)^2}{U_r} = \Sigma\frac{\Delta_r(\eta x) \,.\, \Delta_r(\eta\xi)}{U_r}.$$

Write $h = h_0 + \alpha$, so that α is a quantity of the same order as κ or λ.

For any compartment write

$$P_1 = \Delta\eta\xi,\ P_2 = \Delta\eta\xi^2,\ P_3 = \Delta\eta\xi^3 \ldots\ldots$$
$$Q_1 = \Delta\eta x,\ Q_2 = \Delta\eta x^2,\ Q_3 = \Delta\eta x^3 \ldots\ldots$$
$$A = h_0Q_1 - P_1,\ B = 2h_0Q_2 - P_2,\ C = 3h_0Q_3 - 2P_3,$$
$$F = 4h_0Q_3 - 2P_3,\ G = 12h_0Q_4 - 7P_4,\ H = 9h_0Q_5 - 6P_5.$$

Then if terms containing α^3, κ^3, λ^3, $\alpha^2\kappa$, etc., are neglected, we obtain after reduction

$$\Sigma\frac{E_r^2}{U_r} = \Sigma\frac{1}{U}\{A^2 - 2\kappa AB - 2\lambda AC + \kappa^2(B^2 + 2AF) + 2\kappa\lambda(BC + AG) + \lambda^2(C^2 + 2AH) - 2\alpha\kappa(Q_1B + 2Q_2A) - 2\alpha\lambda(Q_1C + 3Q_2A) + \alpha^2Q_2^2\},$$

the coefficient of α, viz. $2\Sigma\frac{Q_1(Q_1h_0 - P_1)}{U}$, being zero from the equation for h_0.

Σ is to mean that the expression that follows is to be evaluated for each compartment and the results summed.

Express the condition that the partial differential coefficients with respect to α, κ, and λ are zero and we have :

$$\alpha\Sigma\frac{Q_1^2}{U} - \kappa\Sigma\frac{Q_1B + 2Q_2A}{U} - \lambda\Sigma\frac{Q_1C + 3Q_2A}{U} = 0$$
$$-\alpha\Sigma\frac{Q_1B + 2Q_2A}{U} + \kappa\Sigma\frac{B^2 + 2AF}{U} + \lambda\Sigma\frac{BC + AG}{U} = \Sigma\frac{AB}{U}$$
$$-\alpha\Sigma\frac{Q_1C + 3Q_2A}{U} + \kappa\Sigma\frac{BC + AG}{U} + \lambda\Sigma\frac{C^2 + 2AH}{U} = \Sigma\frac{AC}{U}.$$

We have thus a complete solution for the case where κ^2, λ^2 are negligible in comparison with κ, λ.

The writing out is simplified, since we only need the first few decimal places for the quantities, and A is generally small.

For each compartment, after the median is computed, x_r and x_{r+1} are known; ξ_r and ξ_{r+1} are obtained from the tables; Q_1 and P_1 are formed and h computed. Then the P's, Q's, A, B . . . G are computed for each compartment. A specimen of the work is given in (58), *J.S.S.*, 1916, p. 473.

When κ, λ and α are calculated from the above equations $\lambda\xi_r^3 + \kappa\xi_r + \xi_r = x_r(h + \alpha)$ must be solved for each x_r and

$\int_{\xi_r}^{\xi_{r+1}} \frac{1}{\sqrt{\pi}} e^{-\xi^2} d\xi$ compared with the area on $x_r\, x_{r+1}$, in order to test the agreement. The work involves a very great number of systematic arithmetical processes, especially if there are many compartments, but is quite straightforward.

Moderately Abnormal Curves ((57), pp. 415 *seq.*)

When κ^2 and λ^2 are not negligible we can use either the equations connecting them with the moments, or proceeding by selected percentiles we could develop the process of the preceding paragraphs further, but the practical work becomes prohibitive.

As κ and λ increase we reach peculiarities which limit the applicability of the curves; there is a "*break*" if the curve intersects the axis or passes through ∞ and the ordinate becomes negative, and a "*stop*" if there is a minimum ordinate, after which the curve again ascends.

The ordinate of the translated curve is

$$\text{Const.} \times e^{-\xi^2} \div \frac{dx}{d\xi} = \text{Const.} \times e^{-\xi^2} \div a(1 + 2\kappa\xi + 3\lambda\xi^2).$$

There is a break if the denominator becomes negative.

After analysis this reduces to $\kappa^2 < 9(\lambda + \frac{1}{12})^2$ for the region $\xi = -2$ to $\xi = +2$.

There is a stop in the same region if the ordinate has a minimum value. Equating the derived function of the ordinate to zero, we have, when κ^2 and λ^2 are neglected,

$$2\xi(1 + 2\kappa\xi + 3\lambda\xi^2) + 2\kappa + 6\lambda\xi = 0,$$

i.e.
$$3\lambda\xi^3 + 2\kappa\xi^2 + (3\lambda + 1)\xi + \kappa = 0.$$

This is found to have no real root up to $\xi = 2$ other than the mode $\xi = -\kappa$ if $\kappa^2 < \frac{100}{9}(\lambda + \frac{1}{15})^2$.

These conditions are merged in the statement that λ should be above the line *ABD* in the accompanying diagram facing, whose axes show $\chi = \kappa^2$, and λ.

The upper line *EDe* is obtained as follows. If

$$i = \frac{1}{4}\left(\frac{\mu_4}{\mu_2{}^2} - 3\right) = \cdot 3,$$

which gives a rough practical maximum for it, we have from p. 69 above, $2\chi(6\lambda + \frac{75}{4}\lambda^2) = 1 + 5\lambda + \frac{15}{2}\lambda^2 - \frac{45}{4}\lambda^3 - \frac{585}{16}\lambda^4$. It is shown that smaller values of i give a lower curve on the diagram. Hence the irregular triangle *EDBA* shows the limits within which values of λ or χ do not lead to abnormalities.

In the generalised law of error also a break or stop may occur. The former is when

$$1 - 2j(\tau - \tfrac{1}{3}\tau^3) + i(\tfrac{1}{2} - 2\tau^2 + \tfrac{2}{3}\tau^4) = 0$$

where τ is the variable divided by the modulus ((58), *J.S.S.*, 1917, p. 77). The latter is when $i\tau^3 + \frac{4}{3}j\tau^2 + (1 + i)\tau + \frac{2}{3}j = 0$

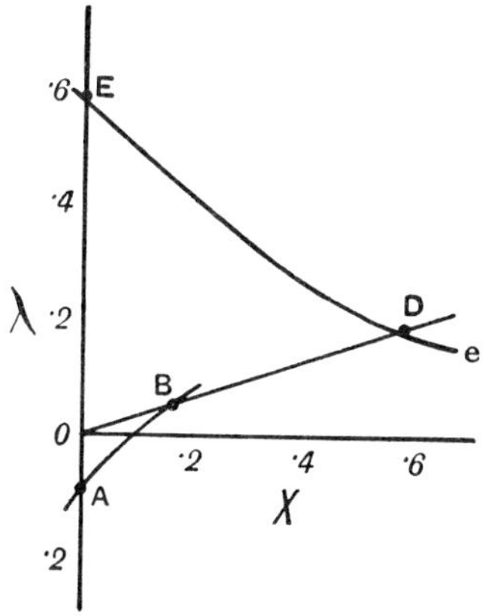

has a root besides $\tau = -\frac{2}{3}j$. It is shown that when j and i are between $\pm$ ·1, no break or stop occurs between the origin and twice the modulus. That is, for subnormal curves we are safe in this respect for about 98% of the observations. For more abnormal curves it is preferable to use the method of translation for all but the central region (cf. p. 63 above).

When κ and λ have greater values than those discussed and we have markedly abnormal curves, the method of translation is still practicable, but the curves lose their typical shape and present singularities, giving J- and U-shaped curves and even curves bent over so that there are two ordinate values to one abscissa. These may in some cases be found to correspond to actual statistics ((57), *J.S.S.*, 1914, p. 653).

As regards the relative merits of the method of translation and the method of the generalised law of error when applied to observational statistics over both the central part and the extremes, Edgeworth's summary is as follows ((58), p. 83): "In the neighbourhood of the normal type either the generalised law or translation or a cross between them may be employed with apparently equal advantage. The choice probably depends upon the purpose in hand. As we recede from the normal type the method of translation becomes preferable. If a single uniform method appropriate to general purposes is required, that method is translation rather than generalisation."

Translation Applied to Correlation ((57), pp. 838 *seq.*). Simple Translation

Consider a generating surface *

$$\zeta = \frac{1}{\pi\sqrt{1-r^2}} e^{-\frac{1}{1-r^2}(\xi_1^2 - 2r\xi_1\xi_2 + \xi_2^2)}$$

and equations of translation

$$x_1 = a_1(\xi_1 + \kappa_1\xi_1^2 + \lambda_1\xi_1^3),$$
$$x_2 = a_2(\xi_2 + \kappa_2\xi_2^2 + \lambda_2\xi_2^3).$$

For x_1 and x_2 separately we have the equations already given (p. 69 above). The origins for x_1 and x_2 are the medians of the observations, and the arithmetic means are at $\frac{1}{2}a_1\kappa_1$, $\frac{1}{2}a_2\kappa_2$. Write R for the coefficient of correlation between x_1 and x_2 calculated from the observations, and ${}_1\mu_2$, ${}_2\mu_2$, for the second moments of x_1 and x_2. Then

$$R\sqrt{{}_1\mu_2 \cdot {}_2\mu_2} = [(x_1 - \tfrac{1}{2}a_1\kappa_1)(x_2 - \tfrac{1}{2}a_2\kappa_2)] \dagger = [x_1x_2] - \tfrac{1}{4}a_1a_2\kappa_1\kappa_2$$
$$= a_1a_2\{[(\xi_1 + \kappa_1\xi_1^2 + \lambda_1\xi_1^3)(\xi_2 + \kappa_2\xi_2^2 + \lambda_2\xi_2^3)] - \tfrac{1}{4}\kappa_1\kappa_2\}.$$

Now $[\xi_1\xi_2] = \frac{r}{2}$, since ξ_1, ξ_2 are in terms of their moduli, if r is their coefficient of correlation;

$$[\xi_1^2\xi_2^2] = \tfrac{1}{4}(1-2r^2);\ [\xi_1^3\xi_2] = \tfrac{3}{4}r = [\xi_1\xi_2^3];\ [\xi_1^3\xi_2^3] = \tfrac{1}{8}(9r+6r^3).$$

These values can be obtained by integration of such expressions as $\int_{-\infty}^{\infty} \xi_1^2\xi_2^2 \cdot \zeta d\xi_1 \cdot d\xi_2$. Where the sum of the indices of

* ξ_1, ξ_2 are used instead of Edgeworth's ξ, η, since η has already been used for another purpose.

† [ ] signifies "the mean of."

ξ_1 and ξ_2 is odd the mean value is zero. Write these values in the equation for R, and we obtain

$$R\sqrt{{}_1\mu_2 \cdot {}_2\mu_2} = a_1a_2\left\{\frac{r}{2} + \frac{3}{4}(\lambda_1 + \lambda_2)r + \frac{1}{2}\kappa_1\kappa_2r^2 + \frac{1}{8}\lambda_1\lambda_2(9r + 6r^3)\right\}.$$

Now $\qquad \sqrt{2{}_1\mu_2} = a_1(1 + \kappa_1^2 + 3\lambda_1 + \frac{15}{4}\lambda_1^2)^{\frac{1}{2}}$

and $\qquad \sqrt{2{}_2\mu_2} = a_2(1 + \kappa_2^2 + 3\lambda_2 + \frac{15}{4}\lambda_2^2)^{\frac{1}{2}}$,

so that a_1a_2 can be eliminated and we have a cubic equation for r in terms of R, κ_1, κ_2, λ_1 and λ_2.

If such terms as κ_1^3, $\kappa_1^2\lambda_1 \ldots \kappa_2^3 \ldots \lambda_2^3$ are neglected, we have

$$R\{1 + \tfrac{1}{2}(\kappa_1^2 + \kappa_2^2) + \tfrac{3}{2}(\lambda_1 + \lambda_2) + \tfrac{3}{4}(\lambda_1^2 + \lambda_2^2) + \tfrac{9}{4}\lambda_1\lambda_2\}$$
$$= r\{1 + \tfrac{3}{2}(\lambda_1 + \lambda_2) + \kappa_1\kappa_2r + \tfrac{9}{4}\lambda_1\lambda_2 + \tfrac{3}{2}\lambda_1\lambda_2r^2\}$$

whence

$$r = R\{1 + \tfrac{3}{4}(\lambda_1^2 + \lambda_2^2) + \tfrac{1}{2}(\kappa_1^2 + \kappa_2^2)\} - \kappa_1\kappa_2R^2 - \tfrac{3}{2}\lambda_1\lambda_2R^3;$$

but it is nearly as simple to solve the complete equation above.

In comparing the observations with the content of the surface, we must consider the element $\zeta\Delta\xi_1 \,.\, \Delta\xi_2$ at ξ_1, ξ_2 to become $z\Delta x_1 \,.\, \Delta x_2$ at x_1, x_2, where z is the ordinate of the translated surface. Here

$$\Delta x_1 \,.\, \Delta x_2 = a_1a_2(1 + 2\kappa_1\xi_1 + 3\lambda_1\xi_1^2)(1 + 2\kappa_2\xi_2 + 3\lambda_2\xi_2^2)\Delta\xi_1\Delta\xi_2$$
$$= w \,.\, \Delta\xi_1\Delta\xi_2, \text{ say.}$$

If ξ_1, ξ_2 and x_1, x_2 are taken as central points of small areas (*e.g.* of the compartments as given in the observations), then $\frac{\zeta}{w}$ gives the estimate for z as a proportion of the total. ξ_1, ξ_2 must be evaluated from the equations connecting them with x_1, x_2, and $\zeta\sqrt{1 - r^2} = \frac{1}{\sqrt{\pi}}e^{-q^2} \,.\, \frac{1}{\sqrt{\pi}}e^{-\xi_2^2}$, where $q(1 - r^2)^{\frac{1}{2}} = \xi_1 - r\xi_2$.

Composite Translation ((58), *J.S.S.*, 1917, pp. 282 *seq.*)

x_1 may depend on ξ_2 as well as on ξ_1, and similarly with x_2. Edgeworth only works out one case, which may be written *

$$x_1 = a_1(\xi_1 + \kappa_1\xi_1^2 + \lambda_1\xi_1^3 + \gamma_1\xi_2^2),$$
$$x_2 = a_2(\xi_2 + \kappa_2\xi_2^2 + \lambda_2\xi_2^3 + \gamma_2\xi_1^2).$$

* Edgeworth omits λ_1, λ_2, but it is as simple to keep them in.

Take the arithmetic means as origins. The means referred to the medians are

$$[x_1] = \frac{a_1}{2}(\kappa_1 + \gamma_1),\ [x_2] = \frac{a_2}{2}(\kappa_2 + \gamma_2).$$

$${}_1\mu_2 = [x_1^2] - [x_1]^2 = \frac{a_1^2}{2}\left(1 + \kappa_1^2 + \gamma_1^2 + 2\kappa_1\gamma_1 r^2 + 3\lambda_1 + \frac{15}{4}\lambda_1^2\right)$$

and $${}_2\mu_2 = \frac{a_2^2}{2}\left(1 + \kappa_2^2 + \gamma_2^2 + 2\kappa_2\gamma_2 r^2 + 3\lambda_2 + \frac{15}{4}\lambda_2^2\right)$$

$$R\sqrt{{}_1\mu_2 \cdot {}_2\mu_2} = [x_1x_2] - \frac{a_1a_2}{4}(\kappa_1 + \gamma_1)(\kappa_2 + \gamma_2)$$

$$= \frac{a_1a_2}{2}\{r + \tfrac{3}{2}(\lambda_1 + \lambda_2)r + (\kappa_1\kappa_2 + \gamma_1\gamma_2)r^2 + \kappa_1\gamma_2 + \kappa_2\gamma_1$$
$$+ \lambda_1\lambda_2(\tfrac{9}{4}r + \tfrac{3}{2}r^3)\}.$$

To obtain γ_1, γ_2 we need higher moments. Write

$$a_1^3k_{30} = \left[x_1 - \frac{a_1}{2}(\kappa_1 + \gamma_1)\right]^3,$$

$$a_1^2a_2k_{21} = \left[\left\{x_1 - \frac{a_1}{2}(\kappa_1 + \gamma_1)\right\}^2\left\{x_2 - \frac{a_2}{2}(\kappa_2 + \gamma_2)\right\}\right],$$

and similarly for k_{12} and k_{03}.

Restrict the investigation to subnormal curves, neglecting κ^2, λ^2, $\kappa\gamma$, etc.

$$k_{30} = \tfrac{9}{4}\kappa_1 + \tfrac{3}{4}\gamma_1(1 + 2r^2) - \tfrac{3}{4}(\kappa_1 + \gamma_1) = \tfrac{3}{2}(\kappa_1 + \gamma_1 r^2);$$

similarly $$k_{03} = \tfrac{3}{2}(\kappa_2 + \gamma_2 r^2).$$

$$k_{21} = \kappa_2\left(\frac{1}{4} + \frac{r^2}{2}\right) + \frac{3}{4}\gamma_2 + \frac{3}{2}(\kappa_1 + \gamma_1)r - \frac{1}{4}(\kappa_2 + \gamma_2) - \frac{1}{2}(\kappa_1 + \gamma_1)r$$
$$= r(\kappa_1 + \gamma_1) + \tfrac{1}{2}(\gamma_2 + \kappa_2 r^2);$$

similarly $$k_{12} = r(\kappa_2 + \gamma_2) + \tfrac{1}{2}(\gamma_1 + \kappa_1 r^2).$$

These four equations give κ_1, κ_2, γ_1, γ_2 in terms of k_{30}, etc., calculated from the observations, and r, which to this order of approximation may be taken as equal to that calculated from the observations.

λ_1 and λ_2, if they are not negligible, must be calculated from the original data, as in the case of simple subnormal curves.

To transfer areas we have $$\Delta x_1\Delta x_2 = \begin{vmatrix} \frac{\partial x_1}{\partial \xi_1} & \frac{\partial x_2}{\partial \xi_1} \\ \frac{\partial x_1}{\partial \xi_2} & \frac{\partial x_2}{\partial \xi_2} \end{vmatrix} \Delta\xi_1\Delta\xi_2$$

$$= \{(1 + 2\kappa_1\xi_1 + 3\lambda_1\xi_1^2)(1 + 2\kappa_2\xi_2 + 3\lambda_2\xi^2) - 4\gamma_1\gamma_2\xi_1\xi_2\}\Delta\xi_1\Delta\xi_2.$$

If $\gamma_1 = \gamma_2 = 0$, we have $k_{21} = \frac{1}{3}(2k_{30}r + k_{03}r^2)$ and $k_{12} = \frac{1}{3}(2k_{03}r + k_{30}r^2)$ from the equation above. If these relations do not hold, simple translation is inadequate and composite translation is necessary.

In all cases dealt with in this Chapter numerical illustrations are given, showing the successful use of the formulæ in a wide variety of frequency groups. In the final paper (71) there is a convincing summary of the use of the method of translation.

Comparison of the Edgeworthian and Pearsonian Systems

Edgeworth always regarded the theoretical statement of the generalised law of error as the natural development of the laws of Gauss, Laplace and Poisson, and, it may be added, as containing the complete statement of Bernoulli's laws. It is a purely abstract law of chance, necessarily obtained when the postulates are satisfied, and independent of any verifications provided by natural frequency groups.

The inquiry how far this law, or the hypothetical variant which leads to the method of translation, does represent observational frequency groups is of quite a different order. If such groups are adequately described (up to the set limits of range) by these curves, defined by arithmetic mean (or median), modulus (or standard deviation) and the further constants j, i or κ, λ (and others for two dimensions), then it is shown that the postulates are not inconsistent with the genesis of such groups. The establishment of this proposition had to Edgeworth a considerable philosophic interest.

On the question of the utility of such detailed description of observational groups, he was always somewhat sceptical. The last words of his elaborate analysis in 1916–17 are, " I submit, too, that the decision turns partly on the purpose to which the representation of statistics is directed. But one of the most difficult questions connected with our investigation is: What is its use? " ((58), *J.S.S.*, 1917, p. 437).

Given that the purpose is " to fit mathematical formulæ to certain kinds of data which are of common occurrence in statistics " ((58), *J.S.S.*, 1916, p. 455), and that the generalised law of error is by no means universal in its applicability, the question turns rather on convenience and success than on assumptions of underlying causes. The Edgeworthian and Pearsonian systems are both based on variants from the law of error; both are definitely known to be the expressions of phenomena which are sometimes present, and both transcend the application to these particular phenomena. Edgeworth supposes that the variant is a convergent linear function of the variable in the normal law; Pearson's equation implies that the modulus (or standard deviation) in that law is replaced by a function of the variable.* For small values of the constants b_1 and b_2 or κ, λ the curves tend to identity. They both depend on the number of observations and first four moments of the group or an equal number of other data. It is not evident *a priori* which is the more suitable. Edgeworth summarises the relative advantages of the two systems as follows ((58), *J.S.S.*, 1917, pp. 412–3).

" A.—*Propositions in favour of the Pearsonian system*

" 1. It has been successfully applied to an immense variety of concrete instances.

" 2. It is demonstrably applicable to every variety of the cases with which we are concerned.

" 3. It is typical of observations formed by the interaction of interdependent (or correlated) elements or causes.

" 4. It elicits from given observations, and enables us to predict with regard to future observations *in pari materia,* the starting-point and finale, the *range,* of a frequency group.

" 5. The constants proper to the formulæ in this system are obtained with peculiar ease from the data.

" 6. The system is distinguished by scientific elegance—the unity of a generic form underlying the variety of the types.

" B.—*Propositions in favour of our system*

" 1. In the extensive class of cases which may be regarded as normal or nearly so, one of the alternative methods, the

* His fundamental equation is $\frac{1}{y}\cdot\frac{dy}{dx} = \frac{x+a}{b_0 + b_1x + b_2x^2 + \ldots}$. The normal law is the simple case where $0 = b_1 = b_2 = \ldots$, and $\frac{1}{y}\cdot\frac{dy}{dx} = \frac{x+a}{-\sigma^2}$, where $-a$ gives the centre.

generalised law of error, has the advantage of being based on a *vera causa*, a law—perhaps the most universal law—of nature; the other alternative, translation, has the advantage of agreeing closely in fact with the true law.

"2. The system is characterised by a simplicity congenial to the purposes of its construction.

"3. It is adapted to the representation of frequency *surfaces*.

"4. It admits of being perfected by the variation of the constants so as to satisfy the criterion of a good fit.

"5. For its application, to calculate the area subtended by a finite portion of the abscissa there is required only a simple table, and one that is ready to hand, that of the probability integral.

"6. It is available where, as not infrequently, the data consist of a few *percentiles*, blocks of area so large as to preclude the accurate ascertainment of moments."

These twelve propositions are discussed in some detail ((58), pp. 413–437).

A. 1. A considerable number of subnormal curves correspond with the Edgeworthian at least as well as with the Pearsonian system. An example is given of a somewhat abnormal curve (ages of brides when the bridegroom is between 24 and 25 years) in which the method of translation is very successfully applied. It is remarked that many more trials have been made of the better known system.

A. 2. The limits to the applicability of the Edgeworthian system are so remote that there are few cases where it breaks down if percentiles are used.

A. 3. Edgeworth did not accept the claim that the Pearsonian equation was typical for the result of interdependent causes, for the ideal experiment on which it was originally based, that of drawing balls from an urn without replacement, is only one of an infinite number of possible examples of inter-dependence; a variation of the experiment, viz. that whenever a black ball is extracted, a white ball is inserted, gives quite a different result ((45), p. 330). In fact, both the Pearsonian equation and the Edgeworthian translation are generally used where the original hypotheses which lead to their statement are discarded. (It may be added that the generalised law of error allows for a certain amount of inter-dependence, and that the distortion represented by translation is of the kind to be expected if that amount is exceeded.)

A. 4. If the limitation of range depends on the hypothesis just named, it has no real support, since the equation represents only an approximation which becomes more and more indefinite towards the extremities. If the limitation is admitted to be the result of "the unsupported substitution of a quadratic function of the variable for a constant in the expression of the normal curve" ((58), *J.S.S.*, 1917, p. 432), then it is on a par with the "stops" and "breaks" (p. 76 above) in the Edgeworthian system.

A. 5. The constants proper to translation would be obtained with equal facility if a more complete table of the values of κ and λ in terms of j and i were constructed once for all.*

A. 6. Equally in translation a generic form underlies a variety of types; but the difference of types is not so important, because one merges into the other without the discontinuity that comes in the Pearsonian system when the factors of the quadratic denominator become imaginary.

B. 1. The theoretical importance of the *a priori* genesis of the generalised law of error has been discussed sufficiently above (pp. 29, 35).

B. 2. The idea of translation is essentially simple; "not dealing with differential equations, it is more accessible to practitioners not conversant with the higher mathematics (p. 434)." †

B. 3. Where simple translation of subnormal surfaces is concerned, translation is straightforward. For abnormal surfaces also it may happen that simple methods yield quite adequate results.

B. 4. The improvement of constants so as to give minimum misfit may theoretically be applied to all systems; but the method of translation and percentiles leads directly to it and suggested the process.

B. 5. When κ and λ are determined, only the normal probability integral is needed to determine the theoretic area, while a table serving the same purpose for the Pearsonian types would

* This has now been done, leading to the table on pp. 123–8 below.

† This statement does not, of course, apply to the more elaborate ways of finding the constants.

be very elaborate. [It should be pointed out that such an elaborate table as that given pp. 123–8 below, or considerable arithmetic labour, is necessary in the earlier stage to determine κ and λ, unless the approximate formulæ can be used, and that a cubic equation must be solved for each value of the abscissa. In the case of subnormal curves, however, $\xi = \frac{x}{a} - \frac{x^2}{a^2}\left(\kappa + \lambda\frac{x}{a}\right)$ is the solution, if κ^2 and λ^2 are neglected; and a good further approximation (neglecting cubes) is

$$\xi = \frac{x}{a} - \frac{x^2}{a^2}\left(\kappa + \lambda\frac{x}{a}\right)\left\{1 - \frac{x}{a}\left(2\kappa + 3\lambda\frac{x}{a}\right)\right\}.$$

The cubic equations, in fact, present little difficulty if only the third decimal place is wanted. Also if κ^2 and λ^2 are negligible, $\kappa = \frac{2}{3}j$, $\lambda = \frac{i}{3} - \frac{8}{9}j^2$; and if κ^2, λ^2 are finite, and κ^3, λ^3 are negligible we have the further direct formulæ on p. 70 above. As a compensation to this, however, we may say that as a matter of practice one of the greatest difficulties of the Pearsonian system is the actual calculation of theoretical frequencies after the constants of the curve have been determined.]

Readers will be surprised to find that simplicity is claimed for the Edgeworthian method. The apparent complexity is due to his continued effort towards refinement. The actual work for a *subnormal* curve, as illustrated on pp. 70–1 above, needs only the calculation of the first four moments, the use of simple straightforward formulæ for the median, κ and λ, the determination of ξ from the first formula given above, and the use of the normal probability integral. If the constants are determined from three selected percentiles (p. 72 above) the work is quite brief.

B. 6. Translation and percentiles afford a straightforward method of dealing with irregular or incomplete data, " a substantive asset, the only one against which I think there is no set off " ((58) p. 437).

CHAPTER VII

APPLICATIONS TO SPECIAL PROBLEMS

NUMEROUS applications of the theory of probability were made by Edgeworth before he had completed his studies of the generalised law of error (47) in (1905) and of the Probable Error of Frequency Constants (51) in (1908). Since some of them depend on the latter work, or rather on the work of earlier writers subsequently criticised and extended, their logical place is after the sections devoted to the general conspectus of his analysis. But a very great part of the applications is based on simpler hypotheses and methods now generally known.

The first important paper on applications of probability to statistics is that read at the Jubilee meeting of the Statistical Society under the title, "Methods of Statistics" (9). The discussion on the paper occupies only nineteen lines of print, of which seven were contributed by the reader of the paper, and the impression obtained is that the fundamental importance and originality of the study were not appreciated. In point of fact for the next fifteen years this paper was the only readily available source from which the methods necessary in the application of mathematical methods to ordinary economic or demographic problems could be drawn. The author's summary (p. 217) is adequate to show his general line of thought. "My principal aim has been the practical application of the Law of Error. The theoretical points which are likely to be least familiar to the reader are: *first*, the proposition that the law of error is equally applicable to the elimination of chance; whether it is, or is not, fulfilled by the observations themselves whose means we are comparing; *secondly*, the relation between the two modes of determining the sought modulus, modes following respectively the analogy of physical errors and of games of chance; the former to be preferred when they differ, the latter important when they agree. No originality is claimed for these principles. The first is the theory of Laplace, the second is the theory of Professor Lexis."

By the "elimination of chance" in this passage is meant the separation of results which might be expected in a random selection from a known population or universe from those which would be obtained only if the selection was not random; for example, whether an observed difference between the mean stature of 2,315 criminals and the mean stature of 8,585 adult males in the general population might be expected in random selections of any 2,315 persons compared with 8,585, or is so great as to be significant of a real defect in criminals' stature. "Consider whether the difference exceeds two or three times the relevant modulus. If it does, the difference is not accidental" (p. 187).

The "modulus" is the quantity c in the equation of the normal curve $y=\frac{1}{c\sqrt{\pi}}e^{-\frac{x^2}{c^2}}$, and is of course the standard deviation of that curve multiplied by $\sqrt{2}$. Edgeworth gives the name *fluctuation* to the quantity c^2.

Laplace's theory is that developed later in the paper on "The Law of Error" (47). Under certain conditions there defined, a series of averages each of n quantities selected from a non-normal curve conforms to a normal curve if n is so great that terms involving $\frac{1}{\sqrt{n}}$ are negligible. It follows that nearly definite chances can be assigned for the occurrence of various deviations. Thus the chance that the deviation should exceed positively or negatively twice the modulus is

$$2\int_2^\infty \frac{1}{\sqrt{\pi}}e^{-x^2}\,dx = \cdot 0047\ldots,$$

a chance that Edgeworth regarded as negligible. It may be added that exactly the same expression would be obtained if terms in $\frac{1}{\sqrt{n}}$ were retained, while those in $\frac{1}{n}$ were neglected (p. 33, l. 3 above), so that great asymmetry in the original "non-normal" curve is permissible for moderate values of n, without sensibly affecting the test.

The "theory of Lexis" is not stated explicitly, but may be regarded as the statement that statistical ratios, such as the ratio of male to total births, $p:(p+q)$, fluctuate normally with a modulus $\sqrt{2pq/n}$, where n is the total number of births under observation. This is the analogy to games of chance; thus the ratio of the number of sixes turned up in n throws of dice to the whole number of throws may be written

$$\frac{1}{6} \pm \sqrt{2 \times \frac{1}{6} \times \frac{5}{6} \times \frac{1}{n}},$$

the term after $\pm$ being the modulus. This method was called "combinatorial" by Lexis. It is contrasted with the "physical" method, where if $x_1, \ldots x_t \ldots x_m$ are m measurements of the same physical quantity and $\bar{x}$ their average, the result may be written $\bar{x} \pm \sqrt{\frac{2\Sigma(x_t - \bar{x})^2}{m}}$, the term after $\pm$ being the modulus. Now if m observations are made each of n births, and the resulting numbers of male births are $p_1 n, p_2 n \ldots p_m n$, then the combinatorial method gives *a priori* that these numbers should conform to a normal curve with centre pn, where p the true proportion tends to equal the average of $p_1 \ldots p_t \ldots p_m$, and with modulus $\sqrt{\frac{2pq}{n}} \times n$, where $q = 1 - p$. But we can tabulate these n numbers *a posteriori* as if they were physical observations and determine the modulus from them as

$$\frac{\sqrt{2\sum_1^n (p_t n - pn)^2}}{m}.$$

Where Lexis's theory applies we ought to have approximately

$$\sqrt{2pqn} = \sqrt{2\sum_1^n (p_t - p)^2 n^2/m}.$$

If this equality is obtained, there is a presumption that the combinatorial method is appropriate; if not, other causes are at work and Edgeworth prefers the latter formula.

In the 1885 paper ((9), p. 182) there is the following classification of the problem: "To find how far the difference between any proposed Means is accidental or indicative of a law? A. When a single mean is compared with a single mean (1) in the case of measurements (α) when the data fulfil the normal law of error, (β) when they do not; (2) in the case of statistical ratios." A(2) is reduced to equivalence to A(1) as explained in the last paragraph. "B. When a set of means is compared with a single mean or a set of means." Since the short discussion of this class does not lead to any important new results, but rather to the correction of methods, which by now are generally known to be erroneous, it is unnecessary to deal further with it. In the class A(1) "it has never been doubted that the modulus must be elicited by an examination of the observations (cannot be predicted from the mere knowledge of their number)" (p. 188). If under (a) the observations are $x_1 \ldots x_t \ldots x_m$ and their average $\bar{x}$, and we write $e_t = \bar{x} \sim x_t$, then the fluctuation or square of the modulus is $c^2 = 2\sum_1^m e_t^2 \div m$ (used rather than $2\sum_1^m e_t^2 \div (m - 1)$ in this paper, pp. 11 and 16 above); this is the most accurate method, but we may take $\Sigma e_t \times \sqrt{\pi}$, or the distance from median to quartile $\div \cdot 477$, as convenient approximations to the modulus, c. Whichever method is used, the modulus for the arithmetic mean of m observations is $c/\sqrt{m}$, and the fluctuation of the mean, c^2/m, is inversely proportional to the number of observations. Under (β), non-normal data, it is necessary to use the formula $c^2 = \sum_1^m e_t^2 \div m$, since the approximate equations for the modulus no longer hold; the fluctuation of the arithmetic mean is still c^2/m and its frequency curve is normal. If we use the median, its fluctuation depends not on c, but on the central ordinate y_0, and equals $\frac{1}{2my_0{}^2}$, when the area of the original frequency curve is reduced to unity; the fluctuation of the median is also normal.

The determination of the modulus in the general case A(1) (β) may present difficulty. In such an example as the fluctuation of the sum of digits, it can be known *a priori*, if all digits from 0 to 9 are equally probable. If the observations are given in the form of a frequency table, and are sufficiently numerous, it can be computed at once from them by the formula

$$c^2 = 2\sum_1^m (\bar{x} - x_t)^2 \div m.$$

It may be known from experience of similar data; *e.g.* many anthropometrical measurements agree in giving the modulus for men's stature as approximately 3·6 inches. But if we are given the mean ages at death in two occupations, or the marriage-rates of two classes, or other means, as contrasted with separate observations, some indirect way of finding the modulus must be devised, such as are instanced in the following paragraphs.

Applications: Modulus determined *a priori*

The modulus being determined for each mean we have the following four cases:

(*a*) A person chooses a suit of cards. Another person guesses what suit it is, with a possibility of thought-transference. In 1,833 trials the number of successful guesses was 510. Here $p = \frac{1}{4}$, $q = \frac{3}{4}$, $n = 1833$; $pn = 458$. If there was no thought-transference the expectation may be written

$$pn \pm \sqrt{2pqn} = 458 \pm 13 \text{ approx.}$$

The actual number of correct guesses was 458 + 52, the difference being four times the modulus. The chance of so great an *excess* * is $\int_4^\infty \frac{1}{\sqrt{\pi}} e^{-x^2} dx < \frac{1}{10^6}$, a chance so small as to make practically certain the existence of some external agency or other interference with the pure chance, $\frac{1}{4}$ at each choice.

(*b*) The mean height of 2,315 criminals differs by 2 inches from the mean height of the general population. The modulus for stature is $\sqrt{13}$ inches. If we take the general population as indefinitely large, so that its mean stature is accurately known, the difference to be expected for the mean of 2,315 persons chosen at random has the modulus $\sqrt{\frac{c^2}{m}} = \sqrt{\frac{13}{2315}} = \cdot 075$. "Thrice this modulus, viz. 0·23, falls far short of the observed difference" (p. 196), viz. 2 inches, and the difference is not accidental.

(*c*) In fact, the mean height of the general population was based on the average of only 8,585 persons, and its "fluctuation" is therefore 13 ÷ 8585. Edgeworth uses the well-known formula that the fluctuation of a difference is the sum of the fluctuations

* The chance of so great a difference, positive or negative, is twice this expression.

of the two means compared, together with the theory that if the means conform to the normal law of error, so does their difference. The modulus for the problem (*b*) therefore becomes

$$\sqrt{(\tfrac{13}{2315} + \tfrac{13}{8585})} = \cdot 085$$

nearly, and the conclusion is as before.

(*d*) Edgeworth takes the ratio of male to female births in English counties as 1041 : 1000. If the combinatorial principle applies the modulus for the number of male births in n births of either sex is $\sqrt{2pqn}$; but Edgeworth uses $p/q = r$ as his variable, and the modulus for r is $(1 + r)\sqrt{\frac{2r}{n}}$, or expressed as per cent. of r is $\frac{100(1+r)}{\sqrt{r}}\sqrt{\frac{2}{n}} = 200\sqrt{\frac{2}{n}}$ when $r = 1{\cdot}041$.*

The modulus for the difference between two proportions p_1, p_2 based respectively on m_1 and m_2 observations is

$$\sqrt{\left(\frac{2p_1q_1}{n_1} + \frac{2p_2q_2}{n_2}\right)}.$$

The modulus for the difference between the corresponding ratios $r_1 = p_1/q_1$, $r_2 = p_2/q_2$ is $\sqrt{\left\{\frac{2(1+r_1)^2r_1}{n_1} + \frac{2(1+r_2)^2r_2}{n_2}\right\}}$. Apply this to the question whether the sex ratio of births in England in the first quarter of the year (1·0404) differs significantly from that in the third quarter (1·0375), the whole number of births per quarter being about 200,000 and the ratios being the averages for the ten years 1870–9. Here the number of births in the 10 quarters on which each mean is based is $n_1 = n_2 = 2{,}000{,}000$; $r_1 = 1{\cdot}0404$, $r_2 = 1{\cdot}0375$. The difference $r_1 - r_2 = \cdot 0029$, and the modulus for this difference is

$$\sqrt{[\{(2{\cdot}0404)^2 \times 1{\cdot}0404 + (2{\cdot}0375)^2 \times 1{\cdot}0375\} \times 2 \div 2{,}000{,}000]} = \cdot 00294.$$

* $\delta(r) = \delta\left(\frac{p}{q}\right) = \frac{q\delta p - p\delta q}{q^2} = \frac{\delta p}{q^2}$, since $p + q = 1$, *and* $\delta q = -\delta p$

$$\therefore \sigma_r^2 = \frac{\sigma_p^2}{q^4} = \frac{pq}{nq^4} = \frac{r(1+r)^2}{n}, \text{ since } p = \frac{r}{1+r} \text{ and } q = \frac{1}{1+r}$$

$$\therefore c_r = \sigma_r\sqrt{2} = (1+r)\sqrt{\frac{2r}{n}} \text{ and } \frac{100c_r}{r} = \frac{100(1+r)}{\sqrt{r}}\sqrt{\frac{2}{n}} = \frac{204{\cdot}1}{1{\cdot}020}\sqrt{\frac{2}{n}} = 200\sqrt{\frac{2}{n}}.$$

Edgeworth apparently gives this as $208\sqrt{\frac{2}{n}}$ (p. 199). He writes $r = 1{\cdot}441$ on p. 198, but this is evidently a slip for 1·041 used on p. 201.

The observed difference is therefore not significant. "For greater security" (p. 201), however, Edgeworth does not use this "combinatorial" method, but calculates the modulus (on the "physical" basis) from the ten observations for each quarter; for the first quarter the modulus so computed is $\frac{\sqrt{44}}{1000} = \cdot 0067$ and for the third quarter $\frac{\sqrt{42}}{1000}$. The moduli for the averages of ten years are therefore $\frac{\sqrt{44}}{1000} \div \sqrt{10}$, since the number of observations is ten, one for each year, and $\frac{\sqrt{42}}{1000} \div \sqrt{10}$, and the modulus for the difference is therefore

$$\sqrt{4{\cdot}4 + 4{\cdot}2} \div 1000 = \cdot 00295,^*$$

which is very nearly equal to the observed difference. If the mean of the third, second and fourth quarter is compared with the third, the observed difference is ·0025, the modulus ·0023.† "There is therefore a slight indication (a probability of about $\frac{8}{9}$) of a real difference in the third quarter" (p. 202).

It is to be noticed that in case (*c*) above that Edgeworth works with the means of the populations only and does not indicate here that, if all the original data were available and used, the formula might have to be modified (see p. 100 below). I have not found that he ever attacked the general problem whether two observed frequency distributions might be expected in two random selections from the same universe, but he may have realised that if the difference between means was significant, then *a fortiori* the universes differed, since he speaks of obtaining the necessary result from only part of the data; I think he did not imply the converse that if the difference between means was insignificant, then the universes were presumably similar, since he only points to absence of evidence of dissimilarity.

* Thus the combinatorial and the physical methods give identical results. Edgeworth does not point this out in this example, but illustrates the correspondence on p. 199.

† $\sqrt{\left(\frac{37{\cdot}7}{30} + \frac{42}{10}\right)} = \sqrt{(1{\cdot}26 + 4{\cdot}2)} = 2{\cdot}3$, all divided by 1000, where the 30 is the number of months averaged in the first, second and fourth quarters, and 37·7 the modulus squared.

Modulus not determined *a priori*

When the combinatorial method is not available we cannot determine the modulus required from the mean alone, but must either know sufficiently the distribution about the mean, or find some external evidence about its fluctuation. One instance has just been given where the fluctuation of birth rates is determined by the variation in 40 three-monthly periods. Another case is in relation to the mean age at death in two occupations; here the modulus is found from the frequency curve showing the relative number of deaths (d_x) at each age in the general population ((9) p. 190). Other methods and generally the disentanglement of moduli in complex problems are illustrated by the following example from the paper (10) "On Methods of Ascertaining Variations in the Rates of Births, Deaths and Marriages," *Statistical Journal*, 1885, in which the examples given in the Jubilee number are elucidated and extended.

We have the following data:

DEATH RATES PER TEN THOUSAND

Columns. Years.	1 1876	2 1877	3 1878	4 1879	5 1880	6 1881	7 1882	8 1883	Sums $\sum_1^8 x$	Means $\bar{x}$	Fluctuations $2\sum_1^8(x-\bar{x})^2 \div 8$
Counties.											
Berks. a .	175	172	187	186	181	153	169	166	1389	$173\frac{1}{2}$	224
Herts. b .	174	165	185	184	176	166*	163	188	1401	169*	175*
Bucks. c .	182	171	186	195	179	162	177	183	1435	$179\frac{1}{2}$	172
Oxford d .	179	182	194	183	180	169	167	166	1420	$177\frac{1}{2}$	162
Beds. e .	196	174	203	195	198	171	181	184	1502	188	246
Cambs. f .	173	177	190	191	187	165	171	181	1435	$179\frac{1}{2}$	158
Sums $\sum_a^f x$.	1079	1041	1145	1134	1101	986	1028	1068	8582	1073	1137*
Means $\bar{x}$.	180	$173\frac{1}{2}$	191	189	$183\frac{1}{2}$	164*	171	178	1430	179	189
Fluctuations $2\sum_a^f(x-\bar{x})^2 \div 6$	124	55	77	50	107	68	73	152	706	88	—

* 186 is given for Herts. 1881, which appears to be a misprint for 166. Some of the figures were worked with 186, others with 166. Those marked * have been amended by using 166 throughout.

Write an entry in, say, the cth row and 4th column as ${}_cx_4$ (= 195). The sum of the first row is $\sum_{t=1}^{t=8} {}_ax_t = 1389$. The average $= 1389 \div 8 = 173\frac{1}{2} = {}_a\bar{x}$. The fluctuation

$$2\sum_{t=1}^{t=8} 2({}_ax_t - {}_a\bar{x})^2 \div 8 = 224.$$

Similarly, the sum of the first column is ${}_a x_1 + \ldots + {}_f x_1 = 1079$. The mean $= 1079 \div 6 = 108 = \bar{x}_1$. The fluctuation

$$= 2\{({}_a x_1 - \bar{x}_1)^2 + \ldots + ({}_f x_1 - \bar{x}_1)^2\} \div 6 = 124.$$

The mean of the horizontal fluctuations (one-sixth of the entries 224,175, etc.) $= 189$ approx. $= C_1^2$, say, and the mean of the vertical fluctuations $= \frac{1}{8}(124 + 55 + \ldots) = 88 = C_2^2$, say.

The fluctuations of the horizontal means

$$= \tfrac{1}{6} \text{ of } \{2({}_a\bar{x} - \bar{x})^2 + 2({}_b\bar{x} - \bar{x})^2 + \ldots\}$$
$$= \tfrac{1}{6} \text{ of } \{2(173\tfrac{1}{2} - 179)^2 + \ldots\} = 46 = C_3^2,$$

say, and the fluctuations of the vertical means

$$= \tfrac{1}{8} \text{ of } \{2(\bar{x}_1 - \bar{x})^2 + 2(\bar{x}_2 - \bar{x})^2 + \ldots\}$$
$$= \tfrac{1}{8} \text{ of } \{2(180 - 179)^2 + \ldots\} = 146 = C_4^2, \text{ say.}$$

Required to determine the modulus to be used in considering whether there is a significant variation in death rates in time or place. If we use the combinatorial method we should take $n = 200{,}000$, the approximate population of each county. The general mean death rate of all the entries is 179 per 10,000, and p, the chance of death of an individual, is ·0179.

$$q = 1 - p = \cdot 9821.$$

Then the modulus for any entry ($p \times 10{,}000$) is

$$\sqrt{\frac{2 \times \cdot 0179 \times \cdot 9821}{200{,}000}} \times 10{,}000 = \sqrt{18},$$

and for the difference between any two entries is

$$\sqrt{18 + 18} = \sqrt{36} = 6;$$

for the difference between any two horizontal averages

$$\sqrt{\tfrac{18}{8} + \tfrac{18}{8}} = \sqrt{4 \cdot 5} = 2 \cdot 1;$$

and between any two vertical averages

$$\sqrt{\tfrac{18}{6} + \tfrac{18}{6}} = \sqrt{6} = 2 \cdot 45.$$

On this basis a great number of the differences are significant.

If, however, we regard the entries as 48 independent measurements of the same quantity, the modulus measured by $\sqrt{\dfrac{2\Sigma(x - \bar{x})^2}{48}}$, where $\bar{x} = 179$, is found to be $\sqrt{230}$,* that is

* Not given by Edgeworth.

three or four times greater than the combinatorial modulus, and if we look at the fluctuations in the rows or columns we find that they are in every case much greater than the 18 given by the combinatorial method. There are therefore influences connected both with the date and with the localities which disperse the rates, and in addition the hypothesis of equal and independent chances in one locality and in one year for all the inhabitants may be inappropriate.

Write C^2 for the sought fluctuation independent of the influence of time and place, appropriate to the death rate in a population of 200,000. Write C_t^2 for the modulus that would be found from the influence of the time factor alone. Then it may be assumed that these are independent and $C^2 + C_t^2 = C_1{}^2 = 189$, the mean of the fluctuations for the six districts over the eight years. If there were no time influence the fluctuation of the means of the columns would be $\frac{C^2}{6}$. Edgeworth argues that there may be correlation between these means and the time factor, so that the fluctuation is not $\frac{C^2}{6} + C_t^2$ but something less, say $\frac{C^2}{6} + aC_t^2 = C_4{}^2 = 146$, where a is less than unity.

Eliminate C_t^2 and we have $C^2 = \dfrac{a189 - 146}{a - \frac{1}{6}} = 52$ when $a = 1$, and less if a is smaller. 52 is therefore a superior limit to C^2. Similarly, write $C_p{}^2$ for the fluctuation in the death rates of localities, due to their special circumstances,* independently of time. We have, similarly, $C^2 + C_p{}^2 = C_2{}^2 = 88$, the mean of the fluctuations for the eight years over the six districts; and $\frac{C^2}{8} + \beta C_p{}^2 = C_3{}^2 = 46$, the fluctuation of the means of the columns, where β is less than unity. Eliminate $C_p{}^2$.

$$C^2 = \frac{\beta 88 - 46}{\beta - \frac{1}{8}} = 48$$

when $\beta = 1$, and this is the maximum value. Other similar

* This may be expressed by saying that the chances of death are p_a, p_b . . . p_f, unequal, and the combinatorial modulus is greater than if they were equal.

tests give in general values near 50 for a superior limit for C^2, while the 18 found by the combinatorial method may be regarded as an inferior limit. Hence when the groups are taken for the same county and same year the modulus is between $\sqrt{50}$ and $\sqrt{18}$, *i.e.* between 7 and 4, and the first entry ${}_ax_1$ may, for example, be written 175 ± 7, or, if the standard deviation is used, 175 ± 5, the numbers 7 or 5 somewhat exaggerating the error to which the death rate regarded as typical of the time and place is subject. Many other examples of the same or similar methods are given (10).

Existence of Trend

The problems of finding whether a series is progressive, whether there is trend in time or other variable, was discussed in 1886 (16). Write $2n + 1$ successive observations as $x_{-n} \ldots x_{-1}$, $x_0, x_1 \ldots x_n$, and let $\bar{x}$ be their average. If the law of progress is not known it is presumably small and an arithmetic progression is a good approximation. Write m for the presumed constant difference per unit time due to the progress. Then the trend line has an equation $x - \bar{x} = mt$, where t has values $-n \ldots -1$, $0, 1 \ldots n$. Using the method of least squares we have to minimise $\sum_{t=-n}^{t=n} (x - \bar{x} - mt)^2$.

$$\text{Hence } m\sum_{-n}^{n} t^2 = \sum_{-n}^{n} tx_t,\text{* (since } \Sigma t = 0)$$

$$= (x_1 - x_{-1}) + 2(x_2 - x_{-2}) + \ldots + n(x_n - x_{-n}).$$

Now consider what value m would have if the x's were (unprogressive) random selections from a normal curve with modulus c. The square of the modulus of any $x_t - x_{-t}$ is $2c^2$, and of $t(x_t - x_{-t}) = 2c^2t^2$. The square of the modulus of the expression $(x_1 - x_{-1}) +$, etc., is therefore $2c^2\sum_{1}^{n} t^2$. Hence the modulus for m is

$$\sqrt{2c^2\sum_{1}^{n} t^2} \div \sum_{-n}^{n} t^2 = c \div \sqrt{2(1^2 + 2^2 + \ldots + n^2)}$$

$$= \frac{c\sqrt{3}}{\sqrt{n(n+1)(2n+1)}}.$$

* This is, of course, the form of the regression coefficient, when x is correlated with t. The method of obtaining it is closely akin to that used six or seven years later for the best value of r (see p. 115 below).

If the computed value of m, viz. $\sum_{1}^{n} t(x_t - x_{-t}) \div 2\sum_{1}^{n} t^2$ (which fluctuates normally *) is more than say twice this expression, there is good evidence of progress. If there are no other data c must be computed from the observations by the formula $c^2 = 2\sum_{-n}^{n}(x_t - \bar{x})^2 \div (2n + 1)$. Where there is significant progress this gives a superior limit for c.

Applications of the use of the modulus and of other methods of the calculus of probability to the elucidation of problems, some practical, others illustrative, are to be found in many of the papers listed and described in the Biography, especially (11), (12), (15), (27), (29), (30), (31), (40), (50), (62), and (64). It is not practicable to summarise them here, and most of them are readily accessible. Each paper has its own flavour of originality, and their very extensive range shows in what a variety of cases methods, to a considerable extent peculiar to Edgeworth's point of view, are applicable to the solution of questions which frequently occur.

* If we do not suppose the x curves to be normal, the frequency of m still tends to normality as n increases.

CHAPTER VIII

THE BEST MEAN

In a group of papers (2), (6), (18), (24), etc., beginning in 1883 and culminating in the "Report of the Committee of the British Association on the Variation in Value of the Monetary Standard" (19), Edgeworth discussed the *raison d'être* and utility of various kinds of means from several points of view. His matured opinions on some of the questions raised are summarised in his article on "The Law of Error" in the *Encyclopædia Britannica*, 1902 (46).

He makes a fundamental distinction between a real or objective mean, and a fictitious or subjective mean. The former is the mean of observations of a physical quantity, observations which are liable to instrumental or other errors; the latter is a position in a statistical frequency group. "The mean of observations is a cause, as it were the source from which diverging errors emanate. The mean of statistics is a description, a representative quantity put for a whole group, the best representative of the group, that quantity which, if we must in practice put one quantity for many, minimises the error unavoidably attending such practice. . . . Observations are different copies of one original; statistics are different originals affording one 'generic portrait'" ((18), pp. 139, 140).

In his principal formal analysis ((18), p. 139) seven other subdivisions of both problems are formally distinguished, making in all a possibility of 256 subspecies. Many of these are shown to be unimportant or even non-existent, and it seems necessary only to discuss those which lead to some characteristic criticism of known methods or to further work in later papers.

Objective Means. The classical problem discussed by Laplace may be thus stated:—A number (n) of observations $x_1 \ldots x_t \ldots x_n$ of some physical object are made by n different instruments, each of which is subject to error grouped in a facility curve $f_t(e)$ with modulus c_t. The best value to take for the measurement of the object is given by the method of least squares as $\sum \frac{x_t}{c_t^2} \div \sum \frac{1}{c_t^2}$. More generally, if the c's are not known, Laplace arrived at the solution that the value was a weighted mean of the x's, the weights being so determined that the fluctuation on the resulting errors should be a minimum.

Edgeworth criticises the procedure on several grounds. (1) The most probable value obtained by the method of genuine inverse probability (p. 16 above) agrees with that just named only if the facility curves are all normal, or all the same. (2) It is tacitly assumed that all values of the objective in the range concerned are equally probable, and the difficulties connected with inverse probability (p. 7 above) are ignored. (3) Even if the solution gives the most probable value, it does not give the most advantageous value if the resultant curve of error is unsymmetrical, for the most advantageous value depends on the unknown quantity of detriment arising from errors of different amounts (p. 11 above). The most advantageous value in cases of (positive) asymmetry is somewhat greater than that given by least squares. The analysis under the Law of Error (p. 43 above) shows that the curve of error may be expected to be unsymmetrical. If there is a zero not far off the practical range, below which the observations cannot fall, the asymmetry is positive; and the same is true, if the deviations are to be regarded as equal if their ratios to the magnitude measured are the same, *e.g.* if an error of one inch in forty is regarded as equal to an error of $1\frac{1}{4}$ inches in 50. (This corresponds to Fechner's law of sensation and to those cases in statistics which lead to the geometric mean.)

Under the first of these criticisms it is remarked ((46), § 26) that the method of least squares is equivalent to operating on the averages of batches of observations, instead of on the individual observations. For averages of batches, selected by definite rules, tend to conform to the normal law of error, where least squares and the genuine inverse method give the same result. Part of the information is thrown overboard in order to give a practical method of procedure. Edgeworth therefore calls least squares the abridged inverse method, especially in the 1908 paper summarised above (pp. 9 *seq*). To prefer practicability to exactness is common in applied statistics, and is not unjustifiable since in the end only a rough measurement of accuracy is possible, owing to the uncertainty of the appropriate hypothesis of inverse probability and to the looseness of the relation between mathematical probability and credibility (p. 7 above). A similar method ((46), § 28) is used in the problem whether the death rates of two classes differ significantly; instead of dealing with the curves of death rates by ages of the classes, the average ages at death are compared,* only this part of the information being used.

Subjective Means. Though the problems of observations and errors on the one hand, and of statistics and deviations on the other, are formally distinct, yet a great part of the analysis relating to them is common to both. Quetelet brought anthropometric and similar statistics into the ambit of the law of error, and vital statistics and many other groups are handled both by the use of the modulus, based on that law, and on Bernoulli's theorems which are closely allied to it. It is not therefore necessary to follow again the arguments relating to least squares, to inverse probability and to credibility, or to develop the differentiation of the problems. The question of the mean rather turns on the rival claims of different means in different cases, and much of the discussion can be found in the papers on index numbers already re-published (Bibliography 19*).

" The excellence of the Mean . . . depends on two qualities : the *precision* with which our Mean of the given samples approximates to that point which, if the series from which the samples are taken were indefinitely prolonged, would constitute a certain assigned Mean of that series, a certain function of the terms; and the representative character of that point " ((18), p. 166). The means to be considered are the arithmetic mean, the greatest ordinate and the median. We may add also the geometric mean and the weighted arithmetic mean from other contexts. Other means " may in general,† in the absence of special reasons

* It is interesting to notice that this suggests the argument developed by Dr. Rhodes (*J.S.S.*, 1926, pp. 544 *seq.*) that when it is considered whether two composite samples arose from the same or different populations, the resulting formula depends on whether we operate on the averages only or use the whole of the original data.

† I do not understand from this that the harmonic mean is always to be rejected, but that it is merely an arithmetic mean of reciprocals. Its use is discussed by Edgeworth in several places.

to the contrary, be rejected on the ground of convenience" ((18), p. 166).

The greatest ordinate of a statistical group has the double disadvantage that its precision has not been computed, and that it has commonly to be determined by a hazardous method of interpolation. It has, however, definite use when part of the frequency group is not known at all—say, observations below a certain minimum in height statistics, where neither mean nor median can be computed.

The geometric mean is definitely appropriate when the *logarithms* of the observations are normally dispersed (as happens in some price index numbers, where the weighted geometric mean is also the most probable ((19*), Vol. I, p. 242) and when we are concerned with relative instead of absolute errors. Since the geometric is less than the arithmetic mean, its use is advisable if errors in defect are less serious than errors in excess ((4), p. 717), and generally it reduces the effect of positive asymmetry which occurs when there is an effective zero of measurement.

The *median* has the disadvantage that its standard deviation of error (1 divided by $2\sqrt{n}$ times the greatest ordinate where the area of the frequency curve is unity) is greater by 25% than that of the arithmetic mean in the normal curve. This excess is not, however, serious in the rough measurements of credibility with which we are generally concerned in statistics, and in some non-normal curves the median is more accurate than the arithmetic mean (see note p. 102 below). The simple median is usually a good approximation to the weighted median; this is advantageous when the weights are not accurately known, or where there is a correlation between the weights and the quantities which disturbs the solution ((19*), p. 331). In positively asymmetrical curves it is less than the arithmetic mean and so has any merits that would accrue to the geometric mean in this case. It has the well-known advantage that it can be computed when the measurements away from the centre are known only roughly, and generally in graded observations interpolation is only needed in the central parts, whereas the arithmetic mean needs correction (by Sheppard's formula or otherwise) for all grades. If no graduation is necessary the median is evidently the easiest mean to compute, and if the maximum ordinate is known its probable error can at once be written down.

As is well known, the median is that position which makes the sum of the deviations from it (all taken as positive) a minimum, the test of least detriment suggested by Laplace. If we take $F(e)$, the measure of the detriment of an error e (p. 10 above), as simply ke, or $k(x-l)$, where x is the variable and l the sought mean, the detriment to be minimised becomes

$$E = \int_l^\infty k(x-l)y\,dx + \int_{-\infty}^l k(l-x)y\,dx \text{ (p. 10 above).}$$

The minimum is given by

$$0 = \frac{dE}{dl} = -\int_l^\infty ky\,dx + \int_{-\infty}^l ky\,dx,$$

that is when the areas to the right and left of l are equal, and l is the median.

If the range of the observations is small, $F(e)$ differs little from $F'(0) \times e$, so that k is identifiable with $F'(0)$.

Finally, if a mean is required of discordant observations, where discordance signifies that the observations are taken from facility curves with different moduli, there is "a peculiar propriety in the use of the Median," as in the example in the footnote * (Bibliography (19*), p. 249).

For these reasons Edgeworth attached great importance to the median in a considerable number of problems. His advocacy extends to its use for two or more unknowns, as explained in the following paragraphs.

* Consider the compound frequency curve

$$y = \frac{1}{2\sqrt{\pi}\,.\,c_1} e^{-\frac{x^2}{c_1^2}} + \frac{1}{2\sqrt{\pi}\,.\,c_2} e^{-\frac{x^2}{c_2^2}}.$$

Write σ_1, σ_2 for the standard deviations of the arithmetic mean and the median.

$$\sigma_1 = \frac{\sqrt{c_1^2 + c_2^2}}{2\sqrt{n}}, \quad \sigma_2 = \frac{c_1 c_2 \sqrt{\pi}}{(c_1 + c_2)\sqrt{n}},$$

since the maximum ordinate is $\frac{1}{2\sqrt{\pi}}\left(\frac{1}{c_1} + \frac{1}{c_2}\right)$, and the standard deviation of the median is $\frac{1}{2y_0\sqrt{n}}$ when y_0 is the maximum ordinate. If $\frac{c_1}{c_2}$ or $\frac{c_2}{c_1} > 2\frac{1}{4}$, $\sigma_1 > \sigma_2$ (Cf. Yule, *Theory of Statistics*, Chap. XVII, § 5).

In relation to the accuracy of the median, an instance is given where in ascertaining whether the difference between the mean age at death of a class differs significantly from that of the general population, the modulus for the difference between the two medians is ·7 of the modulus of the difference between the arithmetic means ((18), p. 169).

The Method of Situation (21), (23), (26), (71)

Suppose that we have n observations of an unknown quantity x, in the form $a_1x - w_1 = 0 \ldots a_tx - w_t = 0 \ldots a_nx - w_n = 0$ where the a's are the known weights to be attached to the observations, the w's are observed values, and

$$\frac{w_1}{a_1} < \frac{w_2}{a_2} < \ldots < \frac{w_n}{a_n}.$$

Required, to minimise $\sum_1^n |(a_tx - w_t)|$, where the vertical lines $|\,|$ signify that $a_tx - w_t$ is to be taken if $a_tx > w_t$ and $w_t - a_tx$ if $w_t > a_tx$.

Add up the a's in order until one, a_r, is found such that $\sum_1^{r-1} a_t < \sum_r^n a_t$ and $\sum_1^r a_t > \sum_{r+1}^n a_t$. Then $x = \frac{w_r}{a_r}$ is the required situation. For the quantity to be minimised may be written

$$f(x) = \sum_1^{r-1} a_t\left(x - \frac{w_t}{a_t}\right) + \sum_{r+1}^n a_t\left(\frac{w_t}{a_t} - x\right) + a_r\left|\left(x - \frac{w_r}{a_r}\right)\right|.$$

Then $$f\left(\frac{w_r}{a_r} + \delta x\right)^* = f\left(\frac{w_r}{a_r}\right) + \delta x\left(\sum_1^r a_t - \sum_{r+1}^n a_t\right) > f\left(\frac{w_r}{a_r}\right)$$

and $$f\left(\frac{w_r}{a_r} - \delta x\right) = f\left(\frac{w_r}{a_r}\right) + \delta x\left(\sum_r^n a_t - \sum_1^{r-1} a_t\right) > f\left(\frac{w_r}{a_r}\right),$$

for in the first case the rth term is $a_r\left(x + \delta x - \frac{w_r}{a_r}\right)$ and in the second $a_r\left\{\frac{w_r}{a_r} - (x - \delta x)\right\}$.

If in adding up the a's we reach $\sum_1^r a_t = \sum_{r+1}^n a_t$, $f(x)$ is a minimum over the range $x = \frac{w_r}{a_r}$ to $\frac{w_{r+1}}{a_{r+1}}$.

The result is called the "weighted median." If $a_1 = a_2 = \ldots = a_n = 1$, we have the unweighted median; here $r = \frac{n+1}{2}$ if n is odd, and the solution is $x = w_{\frac{n+1}{2}}$, while if n is even the solution is anywhere from $w_{\frac{n}{2}}$ to $w_{\frac{n}{2}+1}$.

* $f\left(\frac{w_r}{a_r} + \delta x\right) = f\left(\frac{w_r}{a_r}\right) + \delta x \,.\, f'(x)$, where $\frac{w}{a_r}$ is written for x, neglecting $(\delta x)^2$.

The foregoing paragraphs are substantially Laplace's analysis (*Théorie Analytique des Probabilités*, 1914, Deuxième Appendice, pp. 42 *seq.*), on which Edgeworth bases his development of the method (21).

The investigation can be expressed graphically as follows *: Take for example the equations $x + 3 = 0$, $2x - 2 = 0$, $3x - 6 = 0$, $2x - 6 = 0$, $x - 4 = 0$ arranged so that the values of x $(-3, +1 \ldots)$ are in ascending order; mark them as A_1,

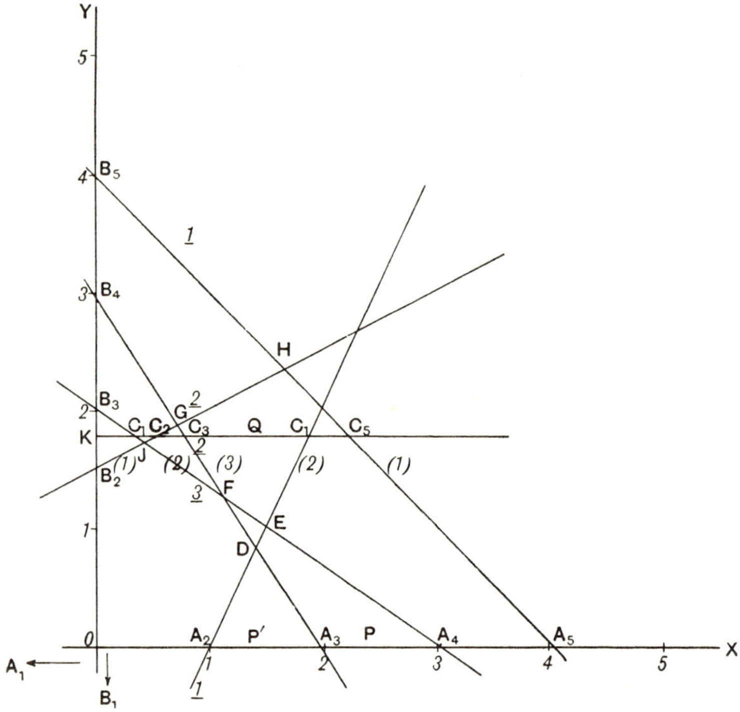

$A_2 \ldots A_5$ on the axis as in the diagram. Then if P is any point on OX, the quantity to be minimised is $PA_1 + 2PA_2 + 3PA_3 + 2PA_4 + PA_5$ each taken positively. The solution is when P is at A_3, for if we move it to the right we add $6A_3P$ to the sum of the first three terms, but subtract only $3A_3P$ from the sum of the last two; while if we move it to the left (P') we subtract $3P'A_3$ from the first two and add $6P'A_3$ to the last three.

Now proceed to the case of two variables x and y given by

* This numerical example is not given by Edgeworth, but seems necessary for clearness.

n equations such as $a_t x + b_t y - w_t = 0$, and investigate what values of x and y make $\sum_1^n |(a_t x + b_t y - w_t)|$ a minimum. The argument is general but can be best followed from a figure. Insert a term in y in the numerical example just given, thus: $x - 2y + 3 = 0, 2x - y - 2 = 0, 3x + 2y - 6 = 0, 2x + 3y - 6 = 0$, $x + y - 4 = 0$. Draw the lines represented by these equations, viz. A_1B_2, A_2B_1, A_3B_4, A_4B_3, A_5B_5, where the B's are the intersections on the axis of y, numbered in the order in which they occur in the figure.

Take any point $Q(x_1y_1)$ for convenience near the centre of the figure, say at $x_1 = 1{\cdot}4$, $y_1 = 1{\cdot}8$. Draw a horizontal line meeting the inclined lines at C_1, C_2 . . ., and OY at K. The horizontal *distance* from any point $Q(x_1, y_1)$ to the intersection C_t with a line $a_t x + b_t y - w_t = 0$ is

$$QC_t = \left| x_1 - \frac{w_t - b_t y_1}{a_t} \right| = \left| \frac{a_t x_1 + b_t y_1 - w_t}{a_t} \right|,$$

where a_t is positive. The quantity to be minimised is therefore $\sum_1^n a_t |QC_t|$.

The problem is that of the weighted median, already solved, when instead of arranging the equations so that the $\frac{w}{a}$'s are in ascending order, we write them so that such terms as $\frac{w_t - b_t y_1}{a_t}$ are in ascending order. The drawing does this automatically, for $KC_t = KQ - C_tQ = x_1 - C_tQ = \frac{w_t - b_t y_1}{a_t}$.

The weight to be given to QC_t is a_t, the coefficient of x in the tth line. These coefficients are put in a row against the lines as (1), (2) . . . . In the position taken for Q the weights come from left to right as 1, 2, 3, 2, 1. Consequently Q must be placed at C_3 to give the minimum when $y = 1{\cdot}8$.

Now vary the value of y and obtain a locus for the weighted median, on which the points C_3 and A_3 (when $y = 0$) are already determined. It is readily seen to be the broken line $A_3DEFC_3GHB_5$.

Next repeat the whole process, interchanging x and y, and find the locus of the weighted median for y when x varies. The equations must be supposed to be re-written so that the coefficients of y are positive. The weights now for the lines are as written vertically on the diagram as *1, 3, 2, 2, 1*. The required locus is B_3JGFEA_4. All points on the broken line EFG are common to the two loci.

Often the solution carried so far, when there are many lines, results in so limited a locus that it is not necessary to proceed further. But if necessary we can determine how the minima for assigned values of x or y change as new values are assigned. Thus in the line GF take a point x', y'. The sum of the weighted horizontal deviations is $(x' - 2y' + 3) + (2x' + 3y' - 6)$ to the left, and $(y' + 2 - 2x') + (4 - y' - x')$ to the right, in total $3 + y'$; that is, the total diminishes with y' and is least at F. Similarly, it is found that the sum of the vertical deviations is $(4 - x' - y') + (x' - 2y' + 3) + (2x' + 3y' - 6) + (y' - 2x' + 2) = 3 + y'$ again and is least at F. Similarly, in the line FE the horizontal minimum at x'', y'' is $3 + x''$ and the vertical minimum is $3 + x''$, again least at F horizontally and vertically. Hence $F(\frac{6}{5}, \frac{6}{5})$ is the only point that satisfies all the conditions, and the absolute minimum of the required expression is 4·2.

The analogy with least squares is thus given: The minimal equations for $\Sigma(ax + by - w)^2 = 0$ are

$$x\Sigma a^2 + y\Sigma ab - \Sigma aw = 0$$
$$x\Sigma ab + y\Sigma b^2 - \Sigma bw = 0.$$

The first equation gives the best value for x for varying values of y, and is analogous to the locus of the x-median. The second equation similarly is analogous to the locus of the y-median. Laplace took the first of these equations and then, substituting the value of x so found in terms of y in the original equations $ax + by - w = 0$, found y by the weighted median method. This procedure suggested to Edgeworth his double median method.

The solutions both by least squares and by the double median method are affected if in any of the original equations the terms are all multiplied by the same constant; thus $x + 2y - 4 = 0$ does not give the same result as $\frac{x}{2} + y - 2 = 0$. It is often somewhat arbitrary which of these forms should be selected. This consideration supports Edgeworth's contention that some indefiniteness in the solution is natural to the problem.

In any case the graphic method is important in that it shows how far the observations are consistent with a unique result, and which of them do not come near the general region of intersection. In the figure the line B_5A_5 represents an erroneous observation.

Edgeworth gives very briefly an analytical solution, of which the foregoing diagram and arithmetic afford an example. He attached so much importance to the method that it is expedient to give the solution in a more general form, and since the only two numerical examples he worked are specially simplified, it is well to add a second illustration, where the method is applied to find the constants in the second approximation to the law of error (pp. 59–60 above).

Write $R = \Sigma \mid a_t x + b_t y - w \mid$.

Take any value x_1, y_1 and determine in which cases $a_t x_1 + b_t y_1 - w$ is positive and in which negative.

Re-write R as $R = \alpha x_1 + \beta y_1 - \gamma - (\alpha' x_1 + \beta' y_1 - \gamma')$, where α, β and γ are the sums of the a's, b's and w's where the deviations are positive, and α', β' and γ' where they are negative.

Now move x_1, y_1 in the direction θ to the axis of x through a distance δs, so that x_1, y_1 become $x_1 + \delta s \,.\, \cos\theta$, $y_1 + \delta s \sin\theta$.

δR, the increment of R, $= \{(\alpha - \alpha')\cos\theta + (\beta - \beta')\sin\theta\}\delta s$.

If x_1, y_1 can be found so that δR is positive for all values of θ, then x_1, y_1 minimise R.

For example take the equations alluded to above (p. 60) in reversed order. Here the coefficients of x are each unity, which simplifies the numerical work but does not affect the generality of the solution.*

The data are

$$\begin{aligned} x - 3{\cdot}58y - 1{\cdot}22 &= 0 \ (a)\\ x - 1{\cdot}88y - 1{\cdot}13 &= 0 \ (b)\\ x - {\cdot}36y - 1{\cdot}57 &= 0 \ (c)\\ x + {\cdot}78y - 1{\cdot}97 &= 0 \ (d)\\ x + 1{\cdot}71y - 2{\cdot}36 &= 0 \ (e)\\ x + 2{\cdot}47y - 2{\cdot}99 &= 0 \ (f) \end{aligned}$$

These are represented by the lines on the figure on next page.

* In the actual example I took the *un*weighted median both for x and y, with (I presume) Edgeworth's approval. It is perhaps a fortunate accident

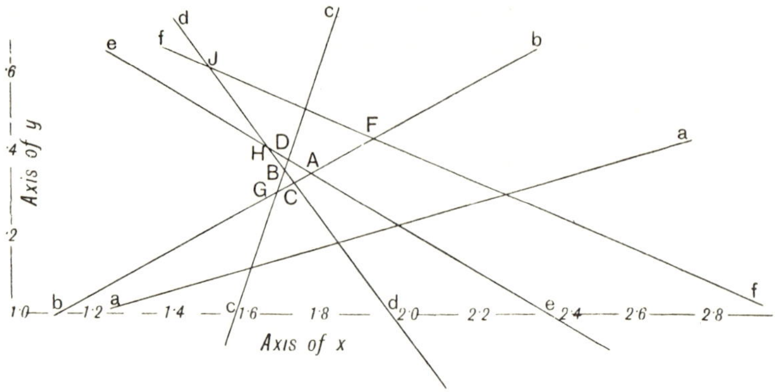

The unweighted median method gives A as the minimising point. Start at the point $C(x_1 = 1{\cdot}72,\ y_1 = {\cdot}31)$, and proceed along the line (b) CA, for which

$$\frac{\sin\theta}{1} = \frac{\cos\theta}{1{\cdot}88} = \frac{1}{\sqrt{1 + 1{\cdot}88^2}} = \frac{1}{2{\cdot}12}.$$

For a point on this line between C and A, $ax + by - w$ is positive for lines (c) and (d) and negative for lines (a), (e) and (f), since a horizontal line cuts the former to the left and the latter to the right. (If we had not all the a's equal to $+ 1$, a rather more elaborate test would be necessary.)

Hence $\alpha = 1 + 1$, $\alpha' = 1 + 1 + 1$, $\beta = - {\cdot}36 + {\cdot}78 = {\cdot}42$, $\beta' = - 3{\cdot}58 + 1{\cdot}71 + 2{\cdot}47 = {\cdot}60$, $\alpha - \alpha' = - 1$, $\beta - \beta' = -{\cdot}18$.

$\delta \mathrm{R} = \delta s(- 1 \times 1{\cdot}88 + {\cdot}18 \times 1) \div 2{\cdot}12 = - {\cdot}97\delta s$ and is negative.

Hence R is diminished as we proceed from C to A.

Now start from A along the line AD (e), for which

$$\frac{\sin\theta}{1} = \frac{\cos\theta}{- 1{\cdot}71} = \frac{1}{1{\cdot}98}.$$

From A to D the horizontal deviations to (c) and (d) are positive and to (a), (b) and (f) are negative. Hence $\alpha - \alpha' = - 1$, $\beta - \beta' = (- {\cdot}36 + {\cdot}78) - (- 3{\cdot}58 - 1{\cdot}88 + 2{\cdot}47) = 3{\cdot}41$;

$\delta R = (- 1 \times - 1{\cdot}71 + 3{\cdot}41 \times 1)\delta s \div 1{\cdot}98 = 2{\cdot}6\delta s$, and is positive.

(illustrating the theorem that the unweighted is a good approximation to the weighted median) that the same point A is obtained by both methods ((45), pp. 341–2).

Now proceed along (*e*) in the opposite direction, away from *D*, so that

$$\frac{\sin\theta}{-1} = \frac{\cos\theta}{1{\cdot}71} = \frac{1}{1{\cdot}98}$$

and deviations to (*b*), (*c*) and (*d*) as positive, and to (*a*) and (*f*) are negative.

$\alpha - \alpha' = 1$, $\beta - \beta' = (-\,1{\cdot}88 - {\cdot}36 + {\cdot}78) - (-\,3{\cdot}58 + 2{\cdot}47)$ $= -\,{\cdot}35$; $\delta R = \{1 \times 1{\cdot}71 + (-\,{\cdot}35) \times (-1)\}\delta s \div 1{\cdot}98 = 1{\cdot}03\delta s$ and is positive.

Finally proceed along *AF*—*C*:

$$\alpha - \alpha' = 1; \;\; \beta - \beta' = (-\,{\cdot}36 + {\cdot}78 + 1{\cdot}71) - (-\,3{\cdot}58 + 2{\cdot}47) = 3{\cdot}24.$$

$\delta R = (1 \times 1{\cdot}88 + 3{\cdot}24)\delta s \div 2{\cdot}12$, and is positive.

Hence if we move from *A* in any one of four inclined directions *R* is increased; *R* is also increased if we move left or right, by taking the simple median, and up or down by taking the weighted median. It can be verified, if it is thought necessary, that the same is true for any intermediate directions.

It is now clear that the full solution must give the position for the weighted median both in x and y. The weighted medians then suggest the final position, and it is only necessary to follow such detail as above when the x and y medians do not give a unique solution, or for verifying the existence of a true minimum.

Edgeworth returned to the subject in 1923 (71), and worked five numerical examples with diagrams. As the first he took the equations above (omitting (*a*)) and re-writing them thus:

$$\begin{aligned}
(b)\;\; & x \times {\cdot}886 - y \times {\cdot}1669 = 1\\
(c)\;\; & x \times {\cdot}635 - y \times {\cdot}2305 = 1\\
(d)\;\; & x \times {\cdot}506 + y \times {\cdot}3935 = 1\\
(e)\;\; & x \times {\cdot}423 + y \times {\cdot}722 = 1\\
(f)\;\; & x \times {\cdot}335 + y \times {\cdot}825 = 1,
\end{aligned}$$

altering the relative weights of the observations.

Using the method of p. 104 above, he finds that the line of the x-medians is *cGCADc′*, where *c′* is on *BD* produced; and the line of the y-medians is *fFADHJf′*, where *f′* is on *FJ* produced. The length *AD* is common to the two loci, and its mid-point is taken as the solution. In fact, however, it can be verified that *D* gives a smaller total than *A*, and that *A* remains the solution.

CHAPTER IX

CORRELATION

EDGEWORTH'S first contribution to the theory of correlation was made in 1892 (32), which was followed by six other papers in 1893 ((33), (34), (35), (36), and (37)) and in 1894 (38). His starting-point was Galton's expression for the frequency of the concurrence of two organs, quoted as

$$\Pi = Je^{-(x_1^2 - 2\rho x_1 x_2 + x_2^2) \div (1-\rho^2)},$$

where the x's were expressed in terms of their quartiles (multiplied by a constant factor), and ρ is an average of the inverse ratios of an x_1 to that value of x_2 which is most frequently associated with it.

This formula he deduced from general principles and extended to any number of variables, reaching the now well-known equation

$z = Ce^{-\frac{1}{D}(x_1^2 R_{11} + x_2^2 R_{22} + \ldots + 2x_1 x_2 R_{12} + \ldots)}$ where $x_1, x_2 \ldots$ are expressed in terms of their moduli, and $R_{11}, R_{12} \ldots$ are the minors of the determinant

$$D = \begin{vmatrix} 1 & \rho_{12} & \rho_{13} & \cdots \\ \rho_{12} & 1 & \rho_{23} & \cdots \\ \vdots & \vdots & \vdots & \end{vmatrix}$$

ρ_{12} is taken as in the previous paragraph, but in 1893 he recommended for its computation the form $\rho_{12} = \dfrac{\Sigma x_1 x_2}{\Sigma x_1^2}$ ((35), p. 101).

The steps in his argument were as follows. It is deduced on general principles * that the frequency of a function of a great number of elements, each varying according to a law of frequency involving x only, is under certain ordinary conditions of the form $He^{-h^2x^2}$ ((33), p. 522). If the laws of frequency of the elements are extended so as to involve y as well as x, then the frequency of the function in x only must be obtainable from that involving x and y by integrating in y (x constant) between its extreme

* See p. 116 below.

limits. Similarly, the function when integrated in x (y constant) must result in the form $Ke^{-k^2y^2}$. "This condition is fulfilled by functions of the form Je^{-R}, where R is a quadratic function of x and y. By parity it may be presumed that, if the law of frequency for each element involve three or more variables, x, y and z, the law of frequency for the compound is Je^{-R}, where R is a quantic of the second order" ((33), p. 523).

For two variables write $R = ax^2 + 2hxy + by^2$, and for any number of variables write

$$R = p_1x_1^2 + p_2x_2^2 + \ldots + 2q_{12}x_1x_2 + \ldots$$

(For three variables write

$$R = ax^2 + by^2 + cz^2 + 2fyz + 2gxz + 2hxy.)$$

Throughout the work suppose the scale for each variable to be so chosen that its modulus is unity.

*[It is implicit in the argument ((32), pp. 198–200) that Galton's form is deducible from this general statement. Write

$$w = Je^{-R} = Je^{-b\left(y+\frac{h}{b}x\right)^2} \cdot e^{-\frac{ab-h^2}{b}x^2}.$$

The frequency of y for any given x is clearly a normal curve with centre at $-\frac{h}{b}x$ and modulus $b^{-\frac{1}{2}}$.

The locus of the y's each of whose frequency is a maximum for the corresponding x is therefore $y = -\frac{h}{b}x = rx$, say, where r is Galton's correlation coefficient. Regression is thus shown to be rectilinear under the appropriate conditions. Similarly, the regression equation for x on y is $x = -\frac{h}{a}y$.

The frequency of x is given by $\int_{-\infty}^{\infty} w\,dy$, x constant during integration, $= J\sqrt{\frac{\pi}{b}} \cdot e^{-\frac{ab-h^2}{b}x^2}$.

Since the modulus is to be unity, $1 = \frac{ab-h^2}{b}$.

Similarly, the consideration of the frequency of y gives $1 = \frac{ab-h^2}{a}$.

* The section marked [ ], ending on p. 112, is not Edgeworth's, but appears to be necessary to clear up his position.

$\therefore a = b$, and the regression equation of x on y is $x = ry$.

Then we have $1 = \dfrac{ab - h^2}{a} = \dfrac{ab - b^2r^2}{a} = a(1 - r^2)$.

Therefore $a = b = \dfrac{1}{1 - r^2}$, and $h = -\dfrac{r}{1 - r^2}$.

J is determined from the equation $\iint z\, dx\, dy = 1$.

Hence we have Galton's equation, with moduli instead of quartile deviations as unity.

Thus no hypothesis, based on observations or otherwise, relating to rectilinearity of regression or the relationship of the two coefficients of regression, is used directly; but the assumption that R is a quadratic function on the preceding page, without consideration of the general form that it may take to satisfy the conditions, involves an undefined hypothesis. This difficulty is cleared up at a later date, especially in (47).]

In fact, Edgeworth first obtains the formula for three variables from the consideration that it should lead to Galton's form when the third variable is integrated away ((32), pp. 194–6).

He then proceeds to re-cast the argument as follows. Write

$$w = Je^{-R}, \text{ where } R = ax^2 + by^2 + cz^2 + 2fyz + 2gxz + 2hxy,$$

and proceed independently of Galton's form. For any given x, say x', the frequency in y, z is represented by a surface whose sections by planes parallel to that of the axes of y, z are similar and concentric ellipses.

The value of y, " to which corresponds the greatest number of instances in the long run," is obtained from the equations $\dfrac{\delta w}{\delta y} = \dfrac{\delta w}{\delta z} = O$ $(x = x')$, which lead to $\dfrac{\delta R}{\delta y} = \dfrac{\delta R}{\delta z} = O$ $(x = x')$, that is $hx' + by + fz = O = gx' + fy + cz$*

$$\therefore y = \frac{fg - ch}{bc - f^2} x' = \frac{H}{A} x',$$

where H, A are minors of the determinant $\Delta = \begin{vmatrix} a & h & g \\ h & b & f \\ g & f & c \end{vmatrix}$.

* In a subsequent note ((34), pp. 63–4) Edgeworth shows that the ratios $a : b : c : f : g : h$ are at once given in terms of the ρ's from these and the two corresponding pairs of equations in y', z' on the assumption that $\rho_{12} = \dfrac{\Sigma y}{\Sigma x'}$, where the most probable y's are summed in the numerator and the corresponding x''s in the denominator. This assumption is justified ((32), p. 200).

Thus $\frac{H}{A}$ is identified with the coefficient of correlation as given by Galton.

Similarly, $z = \frac{G}{A}x'$.

If y' is assigned instead of x', we get similar equations $x = \frac{H}{B}y'$, $z = \frac{F}{B}y'$, and if z' is assigned, $x = \frac{G}{C}z'$, $y = \frac{F}{C}z'$.

Edgeworth proceeds to equate $\frac{H}{A}$ to $\frac{H}{B}$, using Galton's hypothesis, and so obtains $A = B$ and similarly $B = C$. This assumption he implicitly shows to be unnecessary in the following paragraphs.

To find further equations, we have the condition that the complete frequency of x is given by $\int\int w\,dy\,dz$, where x is kept constant during integration.

Write the result of the integration as $J'e^{-dx^2}$, a form which it must take from the considerations in the opening paragraph of the argument (p. 110 above). Now transfer the origin of x to $-\frac{\xi}{d}$, so that we have $J'e^{-d\left(x-\frac{\xi}{d}\right)^2}$, whose centre is $x = \frac{\xi}{d}$. Multiply the earlier expression $\int\int w\,dy\,dz$ equally by $e^{+2\xi x-\frac{\xi^2}{d}}$, so that it becomes

$$\int\int Je^{-\left(ax^2+by^2+cz^2+2fyz+2gxz+2hxy-2\xi x+\frac{\xi^2}{d}\right)}.\,dy\,dz,$$

where it is evident that the new term does not affect the integration.

The greatest frequency for x is then given by differentiating the index successively with respect to x, y, z and equating the results to zero, so that,

$$\begin{aligned} ax + hy + gz &= \xi \\ hx + by + fz &= 0 \\ gx + fy + cz &= 0. \end{aligned}$$

Solving for x we have $x = \frac{A}{\Delta}\xi$ for the ordinate of the centre.

But we already know that the centre is $\frac{\xi}{d}$.

$$\therefore\ d = \frac{\Delta}{A}, \text{ and the modulus for } x \text{ is } \left(\frac{A}{\Delta}\right)^{\frac{1}{2}}.*$$

Now the moduli throughout are to be taken as unity, $\therefore \Delta = A$.

Edgeworth might have proceeded to say, similarly $\Delta = B$ and $\Delta = C$, and so have avoided any dependence on Galton's hypothesis, and, in fact, he gives these equations in another form ((32), p. 200).

Write

$$\frac{H}{A} = \rho_{12} = \frac{H}{B} = \rho_{21};\ \frac{G}{A} = \rho_{13} = \frac{G}{C} = \rho_{31};\ \frac{F}{B} = \rho_{23} = \frac{F}{C} = \rho_{32};$$

Then $\rho_{12} = \rho_{21}$, $\rho_{23} = \rho_{32}$, $\rho_{31} = \rho_{13}$.

Write $$\Delta' = \begin{vmatrix} A & H & G \\ H & B & F \\ G & F & C \end{vmatrix} = \begin{vmatrix} \Delta & \Delta\rho_{12} & \Delta\rho_{13} \\ \Delta\rho_{21} & \Delta & \Delta\rho_{23} \\ \Delta\rho_{31} & \Delta\rho_{32} & \Delta \end{vmatrix} = \Delta^3 D.$$

where $D = 1 + 2\rho_{12}\rho_{23}\rho_{31} - \rho_{12}^2 - \rho_{23}^2 - \rho_{31}^2$.
Then by well-known properties of determinants

$$\Delta a = BC - F^2 = \Delta^2(1 - \rho_{23}^2),$$

$\Delta f = GH - AF = \Delta^2(\rho_{12}\rho_{13} - \rho_{23})$, with similar expressions for b, c, g, h.

Also

$$\Delta = A = bc - f^2 = \Delta^2\{(1 - \rho_{13}^2)(1 - \rho_{12}^2) - (\rho_{12}\rho_{13} - \rho_{23})^2\} = \Delta^2.D$$

$$\therefore\ \Delta = \frac{1}{D},\ a = \frac{1 - \rho_{23}^2}{D},\ f = \frac{\rho_{12}\rho_{13} - \rho_{23}}{D},$$

and the frequency distribution is given by

$$Je^{-\frac{x^2(1-\rho_{23}^2)+y^2(1-\rho_{31}^2)+z^2(1-\rho_{12}^2)+2yz(\rho_{12}\rho_{31}-\rho_{23})+2zx(\rho_{23}\rho_{12}-\rho_{31})+2xy(\rho_{31}\rho_{23}-\rho_{12})}{1+2\rho_{12}\rho_{23}\rho_{31}-\rho_{12}^2-\rho_{23}^2-\rho_{31}^2}}.$$

"This reasoning is quite general; and accordingly . . . we may extend to four and higher numbers of variables the solution which has been given above for the case of three variables." For this purpose the determinant D is extended and the coefficient

* I hope that this is a true elucidation of Edgeworth's argument, which is very condensed and contains apparently at least one misprint. The same result is of course obtainable by direct integration, but this becomes a little intricate as the number of variables is increased. It is so obtained for three variables ((32), p. 195).

of $x_1^2 \ldots x_1x_2 \ldots$ are the minors of D (with appropriate sign) each divided by D.

It is to be observed that in the above analysis the ρ's are rather vaguely connected with the coefficent of correlation. They are implicitly defined as the ratio of the most probable value of (say) y to an assigned value of (say) x, and the analysis shows that this ratio is constant for all values of x. The first method given for calculating r (ρ_{12}) is to divide the sum of all positive values of y by the sum of all the corresponding values (positive or negative) of x ((32), p. 191). But in the following year he took up the question, what is the best value of r? ((35), p. 100).

His argument is as follows when the relevant theorem of least squares is given in detail. Each ratio $\frac{y}{x}$ may be regarded as an observation for r, say $\frac{y}{x} - r = d$. For any y associated with any assigned x the modulus of fluctuation is $\sqrt{1 - r_0^2}$, where r_0 is the true value of r. Hence the modulus for the frequency curve of any $\frac{y}{x}$ is $\frac{\sqrt{1 - r_0^2}}{x} = c$.

The chance of the occurrence of a deviation d is assumed to be $\frac{1}{c\sqrt{\pi}} e^{-\frac{d^2}{c^2}}$, and the chance of the concurrence of all the observed deviations is Constant $\times\, e^{-\Sigma\frac{d^2}{c^2}}$. We have to minimise this index, when the x's and y's are given and the r is the variable.

$$\sum\frac{d^2}{c^2} = \sum\frac{x^2}{1 - r_0^2}\left(\frac{y}{x} - r\right)^2 = \Sigma(y - rx)^2 \div (1 - r_0^2),$$

and is a minimum when $r = \frac{\Sigma xy}{\Sigma x^2}$.

This is when the moduli of x and y are each unity, and leads to $r = \frac{\Sigma xy}{n\sigma_1\sigma_2}$, when $\frac{x}{\sigma_1\sqrt{2}}$, $\frac{y}{\sigma_2\sqrt{2}}$ are written for x and y. It was foreseen that this method was not the best, "but it could not so easily have been foreseen, and is, I think, a curious circumstance, that in seeking the best variety of a method which is not the best I should have been guided to the absolutely best formula" ((51), *J.S.S.*, 1908, p. 395)—that is the formula given by the genuine inverse method.

If the whole argument is now reviewed, it is seen that the equation of the correlation surface is obtained from the theorems that the fluctuation of x independently of y is normal, as is that of y independently of x, while the non-independence of x and y introduces a product term in the index of the frequency equation when x and y both vary, and similarly for several variables. There is no assumption such as that x and y are formed by the addition of normally fluctuating elements, as used by Mr. Elderton in 1906, following Prof. Pearson (*Frequency Curves and Correlation*, p. 109). At the first effort in 1893 Edgeworth arrived at the general equation for the multiple correlation surface based on the most general conceptions. If he had noticed that in his notation $\int\int wxydx\, dy = \frac{1}{2}r$, he would probably have arrived more definitely at the product-sum formula for the correlation coefficient; as it was, he gave it as the " best method."

Note.—The theorem quoted on p. 110 is reached after an argument which is very difficult to follow (33), owing to its compression, its references to previous theorems and to some doubt as to the meaning of the variables. It is not necessary for general purposes to study it, since it is replaced by the general analysis in 1905 (47,) but it is expedient to find its relationship to the conception of correlation or non-independence between elements. I think that the following paragraphs reproduce the gist of the argument.

Let $x_1, x_2 \ . \ . \ . \ x_n$ be a great number of elements subject to different frequency curves.

Write $X = x_1 + x_2 + \ . \ . \ . \ x_n$, where the x's are selected at random independently of each other. Then under certain conditions relating principally to the ranges of the frequency curves, X has normal frequency.

Replace x_1 by $f_1(x_1)$, x_2 by $f_2(x_2) \ . \ . \ .$, and write $x_1 = \bar{x}_1 + \xi_1$, etc., where $\bar{x}_1, \bar{x}_2 \ . \ . \ .$ are the averages of the elemental frequency groups.

Then $X = \bar{X} + \xi_1 f_1'(\bar{x}_1) + \xi_2 f_2'(\bar{x}_2) + \ . \ . \ .$ where $\bar{X}$ is the average of the values of X, and $\xi_1^2, \xi_2^2 \ . \ . \ .$ are neglected.* The $f_1'(\bar{x}_1) \ . \ . \ .$ are constants.

* It is shown in other papers on the law of error that this neglect is justified when n is great, and is not dependent on the rapidity of convergence of each function separately.

If $f_1'(\bar{x}_1), f_2'(\bar{x}_2) \ldots$ are finite and do not differ greatly from each other, x is still normal.

Now regard the x's as small displacements all on the same axis, and suppose that there are also perpendicular displacements $y_1, y_2 \ldots$ such that the concurrence of x_1 and y_1 has a frequency $f_1(x_1, y_1)$.

Write $Y = \bar{Y} + \eta_1 f_1'(\bar{y}_1) + \eta_2 f_2'(\bar{y}_2) + \ldots$
where $\bar{y}_1, \bar{y}_2 \ldots$ are the averages in y and $y_1 = \bar{y}_1 + \eta_1$, etc. Then under the appropriate conditions Y has normal frequency.

Now the complete frequency for x_1, for all values of y_1, is $\int_{l_1}^{l_2} f_1(x_1, y_1) dy_1$, where l_1 and l_2 are extreme values of y_1, and it is, therefore, a function of x_1, say $\phi_1(x_1)$.

Hence X can be expressed as $\bar{X} + \xi_1 \phi_1'(x_1) + \xi_2 \phi_2'(x_1) + \ldots$ and is normally distributed when the y variation is ignored. Similarly, Y is normally distributed when the x variation is ignored. But when we consider the joint frequency of X and Y, the ignoring of the variation of the elemental y's results in the ignoring of Y, and similarly when we interchange the letters Y, y and X, x.

Hence if $F(X, Y)$ represents the joint frequency of X and Y we must have $\int_{-\infty}^{\infty} F(X,Y) dY = He^{-h^2X^2}$, where H, h are constants which determine a normal curve, and similarly

$$\int_{-\infty}^{\infty} F(X,Y) dX = Ke^{-k^2x^2}.$$

Hence $Ce^{-(aX^2 + 2hxyXY + bY^2)}$ is a form for F which satisfies these conditions.

We may interpret X, Y as the measurements of two limbs of an organism, and x_1, y_1, etc., as changes in some variable that affects their size, the variation in x_1 not being independent of the variation of a y_1.

It should be added that the analysis still holds when X is a function, not linear, of $\xi_1, \xi_2 \ldots$, if it satisfies certain conditions.

CHAPTER X

PROBABILITY IN STATISTICS AND UTILITY IN ECONOMICS

THE definite relation found between these two conceptions and between the mathematical methods of handling them is characteristic of Edgeworth's method of thought, and explains in part how he was able to turn from statistics to economics with such great facility. It is akin to the subject of his paper, " Application of Probabilities to Economics " (*Papers Relating to Political Economy*, Vol. II, p. 387), in which he traced many cases where the fundamental ideas of probability were brought into economic theory. The comparison was introduced in 1887, when the Table of Contents of *Metretike* (20) ran : " The Relation of the Calculus of Probabilities to the Philosophy of Utility is partly (I) of Similarity (pp. 1–52); partly (II) of Identity (pp. 52–56)," and was considered again in 1908 in the paper, " On the Probable Errors of Frequency-Constants " (No. 51, pp. 651–661). We may connect it perhaps with the analogy suggested in *Mathematical Psychics*, 1881, p. v, between " the *Principles of Greatest Happiness*, Utilitarian or Egoistic, which constitute the first principles of Ethics and Economics, and those *Principles of Maximum Energy* which are among the highest generalisations of physics, and in virtue of which mathematical reasoning is applicable to physical phenomena quite as complex as human life."

The measurement of belief is as subtle as the measurement of utility. In each case the calculation is inexact and the criteria used are arbitrary. In probability it is often necessary at some stage to assume equal frequency of chances, in the treatment of utility it is assumed that for some purposes all men are equal; many analogies of this kind are treated in *Metretike*, and it is shown that the methods of solution of problems are analogous in the two sciences. Further, both depend on unverified assumptions, the one that experience is the standard of credibility, the other that utility is the standard of good conduct. Both tend to result in non-numerical mathematical statements, that one quantity is equal to, or greater or less than, another, but not how much greater or less. Again, utility governs probability in such questions as, What is the best mean ?, where the detriment of error is considered.

There is close analogy in both the fundamental ideas and in the mathematical expression between the maximising of probability to discover frequency constants and the maximising of utility to determine equilibrium in economics.

In 1908 (51) it is again shown " how widely the subjective element enters into the calculus of probabilities " (p. 651), for the whole must in the end be related to credibility; similarly, utility must be ultimately relative to the feeling of happiness if it is to have an intelligible meaning. In both economics and statistics classical writers seem to have over-estimated the precision of their statements, owing to " undue confidence in untried methods of deduction. . . . There is even a similarity between the particular instruments in each department which have thus proved treacherous—here the postulate of complete independence between events; there, the postulate of perfect competition between persons." " There is common to both studies a certain speculative or dialectical character which recalls the ancient philosophies. ' In wandering mazes lost,' too often both pursue inquiries which seem to practical intellects interminable and uninteresting. These characteristics, it is to be feared, may seem to attach to inquiries, like those pursued in an earlier portion of this Paper, into the *a priori* probabilities of various measurements. In economics, too, other peoples' mathematics are apt to resemble the

' dark-lantern of the spirit
Which none see by but those that bear it.'

" This similarity of complexion is symptomatic of a deeper affinity between the two studies: the peculiarity that they are both in part metaphysical, unlike the recognised mathematical sciences in that they deal with subjective states of consciousness, belief and satisfaction. Jevons, in his powerful plea for the measurement of utility has dwelt on this analogy: ' Previous to the time of Pascal who would have thought of measuring *doubt* and *belief?* ' The successors of Pascal had already before the time of Jevons thought of measuring utility. Bernoulli had initiated the mathematical treatment of advantage, under the head of 'moral expectation.' Bernoulli, in effect, formulated the law of diminishing utility which plays so large a part in the mathematical theory of Political Economy. Laplace employed some of his choicest mathematical reasoning in the evaluation of 'moral expectation.' Even into the theory of physical observations, as we have already seen, Laplace introduces the conception of detriment.

"Among the principles which pertain to the cognate studies of credibility and utility there is an unmistakable resemblance between two which may be summarily expressed as follows. The probability of alternative events, between which, as regards frequency of occurrence, we know of no difference, may, under certain commonly realised conditions, be treated as equal. The utility resulting from the distribution of equal portions of wealth among persons between whose comparative circumstances and dispositions we know of no difference, may, under certain commonly realised conditions, be treated as equal.

"The cognate studies of probability and utility do well to keep together and support each other, for they have not yet won a recognised position among the arts and sciences. They occupy an insecure place between philosophical literature and mathematical physics, frowned on as pedantically precise by the one, and by the other as suspiciously inexact."

Some of these analogies may seem to be far-fetched and fanciful, but there is no doubt that Edgeworth attached great importance to the relationship between the main fields of his study. It was not that he delighted in subtle distinctions for their own sake, though the more obvious never interested him, but that he was always impelled to go back from the superficial to the foundations. It was impossible for him to ignore inverse probability on the ground that the logical treatment was difficult, or to consider probability or utility as numerical measurements without relating them to the psychical concepts of credibility and happiness. It is typical that his first publications * were *New and Old Methods of Ethics*, 1877, and *Mathematical Psychics*, and that his last paper was devoted to a minor question in the fitting of the translated law of error to observa-

* Known to me.

tions. He was lecturer in logic before he was Tooke Professor of Economics and Statistics. Pure abstraction was the primary bent of his mind, but perhaps it was owing to this characteristic that alike in economics and statistics the importance of the results he reached was both practical and permanent.

Table connecting the Constants of Translation with the Moments of Observations

The equation of translation is $x = a(\xi + \kappa\xi^2 + \lambda\xi^3)$.

χ is written for κ^2.

μ_2, μ_3, μ_4 are the second, third and fourth moments of the observations about their average.

$\beta = \mu_3^2 \div \mu_2^3$ and $\epsilon = \frac{1}{12}\mu_4 \div \mu_2^2 - \frac{1}{4}$.

χ and λ are then connected with β and ϵ by the equations.

$$\chi(\tfrac{3}{2} + \chi + 9\lambda + \tfrac{35}{8}\lambda^2)^2 = \tfrac{1}{8}\beta(1 + \chi + 3\lambda + \tfrac{15}{4}\lambda^2)^3.$$

and

$$2\chi + \lambda + \chi^2 + 18\chi\lambda + 9\lambda^2 + 45\chi\lambda^2 + \tfrac{135}{4}\lambda^3 + \tfrac{405}{8}\lambda^4 = \epsilon(1 + \chi + 3\lambda + \tfrac{15}{4}\lambda^2)^2.$$

The table was constructed * by first computing the values of β and ϵ which were given by values of λ and χ at intervals of ·005, within the limits shown in the diagram on p. 77 above; next the values of χ and λ were plotted for the corresponding values of β and ϵ, where β and ϵ were marked on rectangular axes, and two sets of lines were drawn, one set giving the loci of χ for successive constant values of λ, the other the loci of λ for constant χ. From this diagram values of χ and λ were read off for given values of β and ϵ within the range required, and, after smoothing and adjustment by consideration of the differences between successive readings, were entered in the table. The figures in the table are correct to the third decimal place.

* The whole of the calculation was done by Miss F. J. Nicholas at the London School of Economics, enlarging and transforming Mr. F. Brown's table and working (*J.S.S.*, 1924, p. 582).

Table of Values of λ, χ for given β, ϵ.

β.	ϵ.														
	–·050	–·045	–·040	–·035	–·030	–·025	–·020	·015	–·010	–·005	0	·01	·02	·03	·04
0 λ	–·061	–·053	–·046	–·04	–·034	–·028	–·022	–·016	–·010	–·005	0	·010	·019	·028	·036
χ	0	0	0	0	0	0	0	0	0	0	0	0	0	0	0
·01 λ					–·045	–·039	–·033	–·026	–·020	–·014	–·009	·001	·011	·020	·029
χ					·005	·005	·005	·005	·005	·005	·005	·005	·004	·004	·004
·02 λ									–·031	–·025	–·019	–·008	·002	·012	·021
χ									·010	·010	·010	·009	·009	·009	·009
·03 λ												–·018	–·007	·003	·013
χ												·014	·014	·014	·013
·04 λ													–·016	–·006	·004
χ													·019	·019	·018
·05 λ														–·015	–·005
χ														·024	·024

β.	ϵ.														
	·05	·06	·07	·08	·09	·10	·11	·12	·13	·14	·15	·16	·17	·18	·19
0 λ	·044	·052	·060	·067	·074	·081	·088	·095	·101	·108	·115	·121	·127	·133	·139
χ	0	0	0	0	0	0	0	0	0	0	0	0	0	0	0
·01 λ	·037	·045	·053	·061	·068	·075	·082	·089	·096	·103	·110	·116	·122	·128	·134
χ	·004	·004	·004	·004	·004	·004	·004	·004	·004	·004	·004	·004	·004	·004	·004
·02 λ	·030	·038	·046	·054	·062	·069	·077	·084	·091	·097	·104	·111	·117	·123	·130
χ	·008	·008	·008	·008	·008	·008	·008	·008	·007	·007	·007	·007	·007	·007	·007
·03 λ	·022	·031	·039	·048	·056	·063	·071	·078	·085	·092	·099	·106	·112	·118	·125
χ	·013	·013	·012	·012	·012	·012	·012	·012	·011	·011	·011	·011	·011	·011	·011
·04 λ	·014	·023	·032	·041	·049	·057	·064	·072	·079	·086	·093	·100	·107	·113	·120
χ	·018	·018	·017	·017	·017	·016	·016	·016	·016	·015	·015	·015	·015	·015	·015
·05 λ	·005	·015	·025	·034	·042	·050	·058	·066	·073	·080	·088	·095	·101	·108	·114
χ	·023	·023	·022	·021	·021	·020	·020	·020	·020	·019	·019	·019	·019	·018	·018
·06 λ	–·003	·007	·017	·026	·035	·043	·051	·059	·067	·074	·082	·089	·096	·103	·109
χ	·028	·028	·027	·026	·026	·025	·025	·024	·024	·023	·023	·023	·023	·022	·022
·07 λ		–·020	·008	·018	·027	·036	·044	·053	·061	·068	·076	·083	·090	·097	·104
χ		·033	·032	·032	·031	·030	·029	·029	·028	·028	·028	·027	·027	·027	·026
·08 λ			–·002	·009	·019	·028	·037	·046	·054	·062	·070	·077	·084	·092	·099
χ			·038	·037	·036	·035	·034	·034	·033	·032	·032	·031	·031	·031	·031
·09 λ				0	·010	·020	·030	·039	·047	·055	·063	·071	·078	·086	·093
χ				·043	·042	·041	·040	·039	·038	·037	·036	·036	·035	·035	·035
·10 λ					·001	·012	·022	·031	·039	·048	·057	·065	·072	·080	·087
χ					·048	·046	·045	·044	·043	·042	·041	·040	·040	·039	·039
·11 λ							·013	·022	·032	·041	·050	·058	·066	·073	·081
χ							·051	·050	·049	·047	·046	·045	·045	·044	·044
·12 λ								·013	·023	·033	·042	·050	·059	·067	·074
χ								·056	·055	·053	·052	·051	·050	·049	·048
·13 λ									·014	·024	·034	·043	·052	·060	·068
χ									·061	·060	·058	·057	·055	·054	·053
·14 λ										·016	·025	·035	·044	·053	·061
χ										·066	·065	·063	·061	·060	·059
·15 λ												·027	·036	·045	·054
χ												·068	·067	·066	·065
·16 λ													·028	·038	·047
χ													·073	·072	·071
·17 λ													·020	·030	·039
χ													·080	·078	·077
·18 λ															·030
χ															·083

Table of Values of λ, χ for given β, ϵ.

β.		ε.														
		·20	·21	·22	·23	·24	·25	·26	·27	·28	·29	·30	·31	·32	·33	·34
0	λ	·145	·151	·156	·162	·168	·173	·179	·185	·191	·196					
	χ	*0*	*0*	*0*	*0*	*0*	*0*	*0*	*0*	*0*	*0*					
·01	λ	·140	·146	·152	·158	·164	·169	·175	·181	·187	·192	·198				
	χ	*·004*	*·004*	*·004*	*·004*	*·004*	*·004*	*·004*	*·004*	*·004*	*·003*	*·003*				
·02	λ	·136	·142	·148	·154	·160	·165	·171	·177	·183	·188	·193	·199			
	χ	*·007*	*·007*	*·007*	*·007*	*·007*	*·007*	*·007*	*·007*	*·007*	*·007*	*·007*	*·007*			
·03	λ	·131	·137	·143	·149	·155	·161	·167	·173	·178	·183	·189	·195			
	χ	*·011*	*·011*	*·011*	*·011*	*·011*	*·011*	*·011*	*·011*	*·010*	*·010*	*·010*	*·010*			
·04	λ	·126	·132	·139	·145	·151	·157	·163	·168	·174	·179	·185	·191	·197		
	χ	*·015*	*·015*	*·015*	*·014*	*·014*	*·014*	*·014*	*·014*	*·014*	*·014*	*·014*	*·014*	*·014*		
·05	λ	·121	·127	·134	·140	·146	·152	·158	·164	·170	·175	·181	·187	·193	·198	
	χ	*·018*	*·018*	*·018*	*·018*	*·018*	*·018*	*·018*	*·018*	*·017*	*·017*	*·017*	*·017*	*·017*	*·017*	
·06	λ	·116	·122	·129	·135	·141	·147	·153	·159	·165	·171	·177	·183	·188	·194	·200
	χ	*·022*	*·022*	*·022*	*·022*	*·022*	*·022*	*·022*	*·022*	*·021*	*·021*	*·021*	*·021*	*·021*	*·021*	*·021*
·07	λ	·111	·117	·124	·130	·136	·142	·148	·155	·161	·167	·173	·178	·184	·190	·196
	χ	*·026*	*·026*	*·026*	*·025*	*·025*	*·025*	*·025*	*·025*	*·025*	*·025*	*·025*	*·024*	*·024*	*·024*	*·024*
·08	λ	·106	·112	·119	·125	·132	·138	·144	·150	·156	·162	·168	·174	·180	·186	·192
	χ	*·030*	*·030*	*·030*	*·029*	*·029*	*·029*	*·029*	*·028*	*·028*	*·028*	*·028*	*·028*	*·028*	*·028*	*·028*
·09	λ	·100	·107	·113	·120	·127	·133	·139	·145	·152	·158	·163	·170	·176	·182	·188
	χ	*·034*	*·034*	*·034*	*·033*	*·033*	*·033*	*·033*	*·032*	*·032*	*·032*	*·032*	*·032*	*·032*	*·032*	*·032*
·10	λ	·094	·101	·108	·115	·122	·128	·135	·141	·147	·153	·159	·165	·171	·177	·183
	χ	*·038*	*·038*	*·038*	*·037*	*·037*	*·037*	*·037*	*·037*	*·037*	*·036*	*·036*	*·036*	*·036*	*·036*	*·036*
·11	λ	·088	·096	·103	·110	·117	·123	·130	·136	·142	·148	·154	·161	·167	·173	·179
	χ	*·043*	*·043*	*·042*	*·042*	*·042*	*·041*	*·041*	*·041*	*·041*	*·040*	*·040*	*·040*	*·040*	*·039*	*·039*
·12	λ	·082	·090	·097	·104	·111	·118	·125	·132	·138	·144	·150	·157	·163	·169	·175
	χ	*·048*	*·047*	*·047*	*·046*	*·046*	*·045*	*·045*	*·045*	*·045*	*·044*	*·044*	*·044*	*·044*	*·043*	*·043*
·13	λ	·076	·083	·091	·098	·106	·113	·120	·127	·133	·139	·145	·152	·158	·164	·170
	χ	*·053*	*·052*	*·051*	*·051*	*·050*	*·050*	*·050*	*·049*	*·049*	*·049*	*·048*	*·048*	*·048*	*·048*	*·047*
·14	λ	·070	·077	·085	·093	·100	·107	·114	·121	·128	·134	·140	·147	·154	·160	·166
	χ	*·058*	*·057*	*·056*	*·055*	*·055*	*·054*	*·054*	*·053*	*·053*	*·053*	*·052*	*·052*	*·052*	*·052*	*·051*

Table of Values of λ, χ for given β, ϵ.

β.		ϵ.														
		·20	·21	·22	·23	·24	·25	·26	·27	·28	·29	·30	·31	·32	·33	·34
·15	λ	·063	·071	·079	·087	·094	·101	·108	·115	·122	·129	·135	·142	·149	·156	·162
	χ	·063	·062	·061	·060	·060	·059	·058	·058	·057	·057	·056	·056	·056	·056	·055
·16	λ	·055	·064	·072	·080	·088	·096	·103	·110	·117	·124	·130	·137	·144	·151	·157
	χ	·069	·068	·067	·066	·065	·064	·063	·062	·062	·061	·061	·060	·060	·060	·059
·17	λ	·048	·057	·065	·073	·082	·090	·097	·104	·111	·118	·125	·132	·139	·146	·152
	χ	·075	·074	·072	·071	·070	·069	·068	·067	·067	·066	·065	·065	·065	·064	·064
·18	λ	·040	·049	·058	·066	·075	·083	·091	·098	·105	·112	·120	·127	·133	·140	·147
	χ	·082	·080	·078	·077	·076	·075	·074	·073	·072	·071	·070	·070	·069	·069	·068
·19	λ	·031	·040	·050	·059	·068	·076	·084	·092	·099	·107	·114	·121	·128	·135	·142
	χ	·089	·087	·085	·083	·082	·080	·079	·078	·077	·076	·075	·075	·074	·073	·073
·20	λ		·032	·042	·051	·060	·069	·077	·085	·093	·101	·108	·116	·123	·130	·137
	χ		·094	·092	·090	·088	·086	·085	·084	·082	·081	·080	·080	·079	·078	·077
·21	λ				·043	·052	·061	·070	·079	·087	·095	·103	·110	·117	·124	·131
	χ				·097	·095	·093	·091	·090	·088	·087	·086	·085	·084	·083	·082
·22	λ					·045	·054	·063	·072	·080	·088	·097	·104	·111	·119	·126
	χ					·101	·100	·098	·096	·094	·092	·091	·090	·089	·088	·087
·23	λ						·046	·055	·064	·073	·081	·090	·098	·105	·113	·120
	χ						·107	·104	·102	·100	·098	·097	·096	·095	·094	·093
·24	λ							·047	·056	·065	·074	·083	·091	·099	·107	·114
	χ							·111	·109	·107	·105	·103	·102	·101	·099	·098
·25	λ								·048	·057	·066	·076	·084	·092	·100	·108
	χ								·117	·114	·112	·110	·108	·107	·105	·104
·26	λ									·049	·058	·068	·077	·085	·094	·102
	χ									·122	·120	·117	·115	·114	·112	·110
·27	λ										·050	·060	·069	·078	·087	·095
	χ										·128	·125	·123	·120	·118	·116
·28	λ											·052	·062	·071	·080	·088
	χ											·133	·130	·127	·125	·123
·29	λ												·054	·063	·072	·081
	χ												·138	·135	·133	·130
·30	λ													·055	·064	·073
	χ													·144	·141	·138
·31	λ														·055	·065
	χ														·149	·146
·32	λ															·057
	χ															·155

Table of Values of λ, χ for given β, ϵ.

β.	ϵ.														
	·35	·36	·37	·38	·39	·40	·41	·42	·43	·44	·45	·46	·47	·48	·49
·08 λ	·198														
χ	·028														
·09 λ	·193	·199													
χ	·032	·032													
·10 λ	·189	·195													
χ	·035	·035													
·11 λ	·185	·191	·196												
χ	·039	·039	·039												
·12 λ	·181	·187	·192	·198											
χ	·043	·043	·043	·043											
·13 λ	·176	·182	·188	·194	·200										
χ	·047	·047	·047	·047	·047										
·14 λ	·172	·178	·184	·190	·196										
χ	·051	·051	·051	·051	·050										
·15 λ	·168	·174	·180	·186	·192	·197									
χ	·055	·055	·055	·054	·054	·054									
·16 λ	·163	·169	·175	·181	·187	·193	·198								
χ	·059	·059	·059	·058	·058	·058	·058								
·17 λ	·158	·164	·170	·177	·183	·189	·194								
χ	·063	·063	·063	·062	·062	·062	·062								
·18 λ	·154	·160	·166	·172	·178	·184	·190	·197							
χ	·068	·068	·067	·067	·067	·067	·067	·066							
·19 λ	·149	·155	·161	·167	·173	·180	·186	·192	·198						
χ	·072	·072	·072	·071	·071	·071	·071	·070	·070						
·20 λ	·143	·150	·157	·163	·169	·175	·182	·188	·194	·200					
χ	·077	·076	·076	·075	·075	·075	·075	·074	·074	·074					
·21 λ	·138	·145	·152	·158	·164	·170	·177	·183	·189	·195					
χ	·082	·081	·081	·080	·080	·079	·079	·078	·078	·078					
·22 λ	·133	·140	·147	·153	·160	·166	·172	·178	·185	·191	·197				
χ	·087	·086	·085	·085	·084	·084	·083	·083	·082	·082	·082				
·23 λ	·127	·134	·141	·148	·155	·161	·167	·173	·180	·186	·192	·198			
χ	·092	·091	·090	·090	·089	·088	·088	·087	·087	·087	·086	·086			
·24 λ	·121	·129	·136	·143	·150	·156	·163	·169	·175	·182	·188	·194	·200		
χ	·097	·096	·095	·095	·094	·093	·093	·092	·092	·092	·091	·091	·091		

Table of Values of λ, χ for given β, ϵ.

β.	ϵ.														
	·35	·36	·37	·38	·39	·40	·41	·42	·43	·44	·45	·46	·47	·48	·49
·25 λ	·116	·123	·130	·137	·144	·151	·158	·164	·170	·177	·184	·190	·196		
χ	*·103*	*·102*	*·101*	*·100*	*·099*	*·098*	*·097*	*·097*	*·096*	*·096*	*·095*	*·095*	*·095*		
·26 λ	·110	·117	·124	·131	·139	·146	·153	·159	·166	·172	·179	·185	·192	·198	
χ	*·108*	*·107*	*·106*	*·105*	*·104*	*·103*	*·102*	*·102*	*·101*	*·101*	*·100*	*·100*	*·099*	*·099*	
·27 λ	·103	·111	·118	·126	·133	·140	·147	·154	·161	·167	·174	·180	·187	·193	·200
χ	*·114*	*·113*	*·112*	*·111*	*·110*	*·109*	*·108*	*·107*	*·106*	*·106*	*·105*	*·105*	*·104*	*·104*	*·103*
·28 λ	·096	·104	·112	·120	·127	·134	·142	·148	·155	·162	·169	·175	·182	·189	·195
χ	*·121*	*·119*	*·117*	*·116*	*·115*	*·114*	*·113*	*·112*	*·111*	*·111*	*·110*	*·110*	*·109*	*·109*	*·108*
·29 λ	·089	·098	·106	·114	·121	·128	·136	·143	·150	·157	·164	·170	·177	·184	·190
χ	*·128*	*·125*	*·123*	*·122*	*·121*	*·120*	*·119*	*·118*	*·117*	*·116*	*·115*	*·115*	*·114*	*·114*	*·113*
·30 λ	·082	·091	·099	·107	·115	·123	·130	·138	·145	·152	·159	·165	·172	·179	·186
χ	*·135*	*·132*	*·130*	*·129*	*·127*	*·126*	*·125*	*·123*	*·122*	*·121*	*·120*	*·120*	*·119*	*·119*	*·118*
·31 λ	·074	·083	·092	·100	·109	·117	·124	·132	·139	·146	·153	·160	·167	·174	·181
χ	*·143*	*·140*	*·137*	*·135*	*·133*	*·132*	*·130*	*·129*	*·127*	*·126*	*·126*	*·125*	*·124*	*·124*	*·123*
·32 λ	·067	·076	·085	·093	·102	·110	·118	·126	·133	·141	·148	·155	·162	·169	·176
χ	*·151*	*·148*	*·145*	*·142*	*·140*	*·138*	*·136*	*·135*	*·133*	*·132*	*·131*	*·130*	*·129*	*·129*	*·128*
·33 λ		·068	·077	·086	·095	·103	·112	·120	·127	·135	·142	·149	·156	·163	·170
χ		*·157*	*·153*	*·150*	*·148*	*·145*	*·143*	*·142*	*·140*	*·138*	*·137*	*·136*	*·135*	*·134*	*·133*
·34 λ		·060	·069	·078	·087	·096	·105	·113	·121	·129	·136	·143	·151	·158	·165
χ		*·166*	*·162*	*·159*	*·156*	*·153*	*·150*	*·148*	*·146*	*·144*	*·143*	*·142*	*·141*	*·140*	*·139*
·35 λ			·061	·070	·080	·089	·098	·106	·115	·123	·130	·138	·145	·152	·160
χ			*·171*	*·168*	*·165*	*·161*	*·158*	*·155*	*·153*	*·151*	*·149*	*·148*	*·147*	*·145*	*·144*
·36 λ				·062	·072	·081	·090	·099	·108	·116	·124	·132	·140	·147	·154
χ				*·178*	*·174*	*·170*	*·166*	*·163*	*·160*	*·158*	*·156*	*·154*	*·153*	*·151*	*·150*
·37 λ						·073	·082	·092	·101	·109	·118	·126	·134	·141	·148
χ						*·179*	*·175*	*·172*	*·168*	*·166*	*·163*	*·161*	*·160*	*·158*	*·156*
·38 λ						·065	·075	·084	·093	·102	·111	·119	·127	·135	·142
χ						*·189*	*·184*	*·181*	*·177*	*·174*	*·171*	*·169*	*·167*	*·165*	*·163*
·39 λ								·075	·085	·094	·103	·112	·120	·128	·136
χ								*·190*	*·186*	*·182*	*·179*	*·177*	*·174*	*·172*	*·170*
·40 λ									·076	·086	·096	·105	·113	·122	·130
χ									*·196*	*·191*	*·187*	*·185*	*·182*	*·179*	*·177*
·41 λ											·088	·097	·106	·115	·123
χ											*·196*	*·193*	*·190*	*·187*	*·185*
·42 λ													·099	·107	·116
χ													*·199*	*·196*	*·193*

Table of Values of λ, χ for given β, ϵ.

β.	ϵ.														
	·50	·51	·52	·53	·54	·55	·56	·57	·58	·59	·60	·61	·62	·63	·64
·29 λ	·196														
χ	*·113*														
·30 λ	·192	·198													
χ	*·117*	*·117*													
·31 λ	·187	·194	·200												
χ	*·122*	*·122*	*·121*												
·32 λ	·182	·189	·195												
χ	*·127*	*·126*	*·125*												
·33 λ	·177	·184	·190	·197											
χ	*·132*	*·131*	*·130*	*·130*											
·34 λ	·172	·179	·185	·192	·199										
χ	*·138*	*·137*	*·136*	*·135*	*·135*										
·35 λ	·167	·174	·180	·187	·194										
χ	*·143*	*·142*	*·141*	*·140*	*·140*										
·36 λ	·161	·168	·175	·182	·189	·195									
χ	*·149*	*·148*	*·147*	*·146*	*·145*	*·144*									
·37 λ	·156	·163	·170	·177	·184	·191	·198								
χ	*·155*	*·154*	*·153*	*·152*	*·151*	*·150*	*·149*								
·38 λ	·150	·157	·165	·172	·179	·186	·192	·199							
χ	*·161*	*·160*	*·159*	*·158*	*·157*	*·156*	*·155*	*·154*							
·39 λ	·144	·152	·159	·167	·174	·181	·187	·194							
χ	*·168*	*·166*	*·165*	*·164*	*·162*	*·161*	*·160*	*·159*							
·40 λ	·138	·146	·153	·161	·168	·175	·182	·189	·196						
χ	*·175*	*·173*	*·171*	*·170*	*·168*	*·167*	*·166*	*·165*	*·164*						
·41 λ	·132	·140	·147	·155	·163	·170	·177	·184	·191	·198					
χ	*·182*	*·180*	*·178*	*·176*	*·174*	*·173*	*·172*	*·171*	*·169*	*·169*					
·42 λ	·125	·133	·141	·149	·157	·164	·171	·178	·185	·192	·199				
χ	*·190*	*·187*	*·185*	*·183*	*·181*	*·180*	*·178*	*·177*	*·175*	*·174*	*·173*				
·43 λ	·118	·127	·135	·143	·150	·158	·166	·173	·180	·187	·194				
χ	*·198*	*·195*	*·193*	*·190*	*·188*	*·186*	*·184*	*·183*	*·182*	*·180*	*·179*				
·44 λ				·136	·144	·152	·160	·167	·175	·182	·189	·196			
χ				*·198*	*·195*	*·193*	*·191*	*·190*	*·188*	*·187*	*·186*	*·185*			
·45 λ							·153	·161	·169	·176	·183	·190	·197		
χ							*·199*	*·197*	*·195*	*·193*	*·192*	*·191*	*·190*		
·46 λ										·170	·177	·184	·191	·198	·205
χ										*·200*	*·199*	*·198*	*·197*	*·195*	*·194*

ANNOTATED BIBLIOGRAPHY

Contributions to Mathematical Statistics and the Theory of Probability by Professor F. Y. Edgeworth in approximate order of date of publication.

(Abbreviations used : *Phil. Mag.* for *London, Edinburgh and Dublin Philosophical Magazine*, Vth Series, 1876–1900, and the VIth Series, 1901–1925 (each 50 volumes).

J.S.S. for *Journal of the Royal Statistical Society.*)

NOTE.—With the exception of No. 19,* articles published in the three volumes of Edgeworth's *Papers Relating to Political Economy* are not here included.

1. The Law of Error. *Phil. Mag.* 1883, Vol. XVI, pp. 300–309.

First approach to the 1905 paper (No. 47 below). The law of error is not perfectly attained, " but if the extravagant pretensions to numerical accuracy be abandoned, the moderate claim put in for mathematical reasoning may be received with more favour " (p. 309).

2. The Method of Least Squares. *Phil. Mag.* 1883, Vol. XVI, pp. 360–375.

The subject of *Metretike* (Appendix) (No. 20 below) introduced. Least squares give maximum probability but not necessarily maximum advantage. Measurement of detriment discussed; neither the minimising of mean square deviation nor of mean deviation are logical, but an unknown function of the error measures detriment. See text, p. 10 above.

3. The Physical Basis of Probability. *Phil. Mag.* 1883, Vol. XVI, pp. 433–435.

The formulæ representing energy and its minimum in physics are akin to those representing minimum disadvantage in choice of a mean in probability. The principle of maximising kinetic energy is akin to that of maximising satisfaction in economics.

4. On the Method of ascertaining a Change in the Value of Gold. *J.S.S.* 1883, pp. 714–718.

First essay towards the general problem of index numbers of prices, subsequently developed for the British Association Committee (No. 19 below). Comparison of the uses of the geometric and arithmetic means, and defence of approximation in weighting.

5. The Philosophy of Chance. *Mind*, 1884, pp. 223–235.

Discussed in text, pp. 6 *seq*. above.

6. On the Reduction of Observations. *Phil. Mag.* 1884, Vol. XVII, pp. 135–141.

Continues the subject of No. 2 above, and develops the question whether it is possible to determine what function of the observations is absolutely the most advantageous. Cf. No. 18 below.

7. *A priori* Probabilities. *Phil Mag.* 1884, Vol. XVIII, pp. 204–210.

Discussed in text, pp. 7 *seq*. above.

8. Chance and Law. *Hermathena. A series of papers on Literature, Science and Philosophy by Members of Trinity College, Dublin.* (Dublin), 1884, No. X. (Pp. 154–163 in Vol V, 1885, which contains No. X.)

Examination of the question, "Does the calculus of probabilities afford any argument against or for the existence of chance in its old sense of immunity from scientific law?"

9. Methods of Statistics (with discussion). *Jubilee Volume of the Statistical Society*, 1885, pp. 181–217.

The application of the law of error and in particular the modulus to statistical problems. See text, pp. 86 *seq*. above.

10. On Methods of Ascertaining Variations in the Rate of Births, Deaths and Marriages. *J.S.S.* 1885, pp. 628–652.

Detailed discussion of some problems introduced in No. 9 above. See text, pp. 93 *seq*. above. The methods are original and very ingenious.

11. The Calculus of Probabilities Applied to Psychical Research. *Proceedings of the Society for Psychical Research*, Vol. III, 1885, pp. 190–199.

Application of the theory of probability to test the existence of some agency other than chance in simple experiments in thought transference.

12. The Calculus of Probabilities Applied to Psychical Research, II. *Proceedings of the Society for Psychical Research*, Vol. IV, 1886, pp. 189–208.

Alternative methods are given, which involve some development of the law of error (as in No. 13 below), to fit different cases (see also Nos. 20 and 62 below).

13. The Law of Error and Elimination of Chance. *Phil. Mag.* 1886, Vol. XXI, pp. 308–324.

The problem is to determine whether a divergence is due to chance or disturbing causes. Special analysis is given of the range over which the normal curve is applicable, and whether it affords a safe answer to the problem for divergences beyond that range.

14. On the Determination of the Modulus of Errors. *Phil. Mag.* 1886, Vol. XXI, pp. 500–507.

Where the modulus is not exactly determinable owing to roughness of grading or other causes, a superior limit can sometimes be found. The formulæ given have some analogy to Dr. Sheppard's corrections.

15. Problems in Probabilities. *Phil. Mag.* 1886, Vol. XXII, pp. 371–384.

Given a record of the bank notes in the hands of the public over a series of dates, to find the probability that in the future the number will not exceed certain limits.

16. Progressive Means. *J.S.S.* 1886, pp. 469–475. (Read to the British Association, October, 1885).

Method of testing whether a series (in time or any other variable) fluctuates about a steady average or progresses. See text, p. 96 above.

17. The Mathematical Method of Statistics. *J.S.S.* 1886, pp. 649–654.

Summarises papers in *Phil. Mag.*, Nos. 13 and 14 above.

18. Observations and Statistics: An Essay on the Theory of Errors of Observation and the First Principles of Statistics. (Read May 25th, 1885.) *Transactions of the Cambridge Philosophical Society*, 1887, Vol. XIV, Part II, pp. 138–169.

The paper distinguishes and examines the different cases which are presented by the problem, "What is the best mean?" It is not possible to summarise it. Two of the principal topics, namely the validity of least squares and the measurement of the detriment of error, are dealt with in the text (pp. 10 *seq.* above). The ideas in relation to means in earlier papers are developed, and the whole marks an important stage in the treatment of the subject and is frequently referred to in subsequent papers.

19. (Four) Reports of the Committee appointed for the purpose of investigating the Best Methods of ascertaining and measuring Variation in the Value of the Monetary Standard (Mr. F. Y. Edgeworth, secretary), in the volume of the *Reports of the British Association for the Advancement of Science*:

1887, p. 247, First Report of the Committee, drawn up by the Secretary.

1887, pp. 254–301, Memorandum by the Secretary.

1888, p. 181, Second Report, drawn up by Mr. Giffen.

1888, pp. 188–219, Memorandum by the Secretary.

1889, p. 133, Third Report (merely formal, recommending re-appointment).

1889, pp. 133–164, Memorandum by the Secretary.

1890, p. 485, Fourth Report. (No memorandum.)

19*. The Memoranda are reprinted with critical introduction in Edgeworth's *Papers relating to Political Economy*, Vol. I, pp. 195 *seq.*, and are therefore not summarised in the text.

Apart from the theory of index numbers, these Memoranda contain very important discussions on the general problems of weighted averages.

20. *Metretike or the Method of Measuring Probabilities and Utility.* London: The Temple Company, 1887.

Summarised in text (p. 118 above) as regards utility. In the Appendix especially the validity of the method of least squares is discussed; this and the treatment of *a priori* probability and measurement of detriment are repeated in later papers and summarised in the text above (*passim*).

21. On Observations relating to Several Quantities. *Hermathena*, 1887, No XIII (pp. 279–285 in Vol. VI, 1888, which contains No. XIII).

Summarised in text, pp. 103 *seq.* above.

22. On Discordant Observations. *Phil. Mag.* 1887, Vol. XXIII, pp. 364—375.

How to distinguish errors of observation proper from mistakes. Various criteria for rejection of extreme observations (methods of Airy, Stone, Chauvenet, Peirce, de Morgan, Glaisher and Newcomb).

23. Letter calling attention to article in *Hermathena* (No. 21 above). *Phil. Mag.* 1887, Vol. XXIV, pp. 222–223.

24. The Choice of Means. *Phil. Mag.* 1887, Vol. XXIV, pp. 268–271.

The median contrasted with the result of the method of least squares in the case of a frequency curve which is compounded of different probability curves.

25. Empirical Proof of Law of Error. *Phil. Mag.* 1887, Vol. XXIV, pp. 330–342.

Comparison of the fitness of the law of error with that of some other possible equations to represent the frequency of observations, tested by certain statistics, with verdict in favour of the law of error.

26. On a New Method of Reducing Observations Relating to Several Quantities. *Phil. Mag.* 1888, Vol. XXV, pp. 184–191.

The multiple median method (see text, pp. 103–9 above and (71) below).

27. The Mathematical Theory of Banking. *J.S.S.* 1888, pp. 113–127.

The subject of No. 15 above further treated. Ends with quotation from *Nightmare Abbey* :—" Madam, I should lose my transcendental reputation, if it could be said of Ferdinando Flosky that he was ever known to give any useful information to any one." Edgeworth continues : " The theorist must not pretend to wisdom, if he knows so little what he is about as to mistake his abstract formulæ for rules immediately applicable to practice."

28. Some New Methods of Measuring Variation in General Prices. *J.S.S.* 1888, pp. 346–368.

The weighted median is recommended for some aspects of the problem, but for measuring change in the quantity of gold in relation to its work a weighted average should be used.

29. The Statistics of Examinations. *J.S.S.* 1888, pp. 599–635.

The variation in marking due to various causes is analysed, including the variation of the candidate's performance in relation to his true level. Placing in classes is to be preferred to placing in order.

30. The Element of Chance in Competitive Examinations. *J.S.S.* 1890, pp. 460–475 and 644–663.

It is shown that there is considerable variation in examiners' marking and the element of chance in the resulting order of candidates is computed. It is recommended that the law of error should be used in regulating marks, a procedure subsequently adopted in certain public examinations.

31. Problems in Probabilities. No. 2 : Competitive Examinations. *Phil. Mag.* 1890, Vol. XXX, pp. 171–188.

Mainly a mathematical *résumé* of No. 30.

32. Correlated Averages. *Phil. Mag.* 1892, Vol. XXXIV, pp. 190–204.

33. The Law of Error and Correlated Averages. *Id.* pp. 429–438 and 518–526.

34. A New Method of treating Correlated Averages. *Phil. Mag.* 1893, Vol. XXXV, pp. 63–64.

35. Exercises in the Calculation of Errors. *Phil. Mag.* 1893, Vol. XXXVI, pp. 98–111.

36. Note on the Calculation of Correlation between Organs. *Id.* pp. 350–351.

For a general account of Nos. 32–36 see text, Chapter IX above. Together they show the great advance in the theory of correlation that is attributable to their author.

37. Statistical Correlation between Social Phenomena. *J.S.S.* 1893, pp. 670–675.

Contains the answer to Cournot's objection that Quetelet's "mean man," constructed by the means of separate limbs, would be an impossible monster; the theory of correlation demonstrates the contrary.

38. Asymmetrical Correlation between Social Phenomena. *J.S.S.* 1894, pp. 563–568.

If the frequency curves in x and y are asymmetrical, the correlation surface is distorted and regression is parabolic or even more distorted. Illustration by the statistics of ages of bride and bridegroom.

39. On Some Recent Contributions to the Theory of Statistics. *J.S.S.* 1895, pp. 506–515.

Edgeworth's first reflections on Prof. Pearson's general system of frequency curves. He points out that the basis of the system is not sufficiently general to represent the result of plurality of independent causes except in particular cases and argues that the "distinction between curves with limited and unlimited range does not correspond to any real difference." "It is a nice question what weight should be attached to the correspondence [between the formulæ and the facts] by one who does not perceive any theoretical reason for the formulæ." (p. 511).

40. Statistics of Unprogressive Communities, and Supplementary Notes on Statistics. *J.S.S.* 1896, pp. 358–386 and 529–539.

The Application of the modulus to the question of the mean time of a wasp's flight, as an example of observations on society free from the disturbing influence of time, and also an illustration of the stability of the modulus. The subject is taken up again in Nos. 43, 50 and 66 below. The supplementary notes deal not only with wasps, but continue No. 39 above, refer to Pareto's formula and to the coefficient of correlation.

41. The Asymmetrical Probability-Curve. *Phil. Mag.* 1896, Vol. XLI, pp. 90–99.

Various proofs of the equation to the second approximation to the law of error.

42. The Compound Law of Error. *Phil. Mag.* 1896, Vol. XLI, pp. 207–215.

The generalised normal correlation surface is obtained briefly by a system of partial differential equations, and is extended for two variables to include unsymmetrical curves. The results and methods of Nos. 41 and 42 are included in No. 47 below.

43. Miscellaneous Applications of the Calculus of Probabilities. *J.S.S.* 1897, pp. 681–698 and 1898, pp. 119–131 and 534–544.

Smallness of improvement in stability that would be made by bimetallism. Stability of duration of voyage of bees. Is excess of male over female births really greater in some districts than in others? Does the system of voting result in the proportions of parties in Parliament reproducing closely the proportions in the electorate?

44. On the Representation of Statistics by Mathematical Formulæ. Part I, *J.S.S.* 1898, pp. 670–700. Part II, *J.S.S.* 1899, pp. 125–140. Part III, *J.S.S.* 1899, pp. 373–385. Part IV, *J.S.S.* 1899, pp. 534–555. Supplement, *J.S.S.* 1900, pp. 72–81.

These (without the Supplement) were issued in a volume for private circulation together with xix pages of Introductory Description and 1 page of Corrigenda et Addenda. See text, pp. 66 *seq.* above.

45. Methods of Representing Statistics of Wages and other Groups not fulfilling the Normal Law of Error (with A. L. Bowley). *J.S.S.* 1902, pp. 325–354.

See text, pp. 56–60 above.

46. Law of Error. *Encyc. Brit.* Xth Edition, 1902, Vol. XXVIII, pp. 280–291.

A general analysis containing little that is new to readers of the former papers, but summarising them in a convenient form.

47. The Law of Error. *Cambridge Philosophical Transactions*, 1905, Vol. XX, Part I, pp. 36–141. (Received Agust, 1904.) Circulated privately with unpublished Continuation of Appendix VII, pp. i–xiv.

The developed result of the series of investigations, such as in Nos. 1, 13, 25, 33, 41 and 42, of the frequency of quantities whose magnitudes depend on a complex of fluctuating elements and of the conditions which lead to a general law (see pp. 29–50 above). The most important of the papers relating to this subject.

48. The Generalised Law of Error, or the Law of Great Numbers. *J.S.S.* 1906, pp. 497–530, with discussion pp. 531–9.

A summary of No. 47 with explanations and some further analysis.

49. On the Representation of Statistical Frequency by a Series. *J.S.S.* 1907, pp. 102–106.

Criticism of the Bruns-Charlier series. It is not the true generalisation of Laplace's law of error, it does not satisfy the condition that the inclusion of a new element should not alter the form, nor is it reproductive in the sense that if P and Q fulfil the law, then $R = P + Q$ fulfils it; nor are mean powers of the curve and of the observations consilient beyond the second approximation. In all these respects it is inferior to the generalised law of error (No. 47 above). Exception is taken to the misunderstanding of the basis of the generalised law when it is confused with the Charlier series (*Frequency Curves and Correlation*, by W. P. Elderton, 1906, pp. 161–162). See foot-note, p. 35 above.

50. Statistical Observations on Wasps and Bees. *Biometrika*, 1907, Vol. V, No. 4, pp. 365–386.

See No. 40 above and No. 66 below.

51. On the Probable Errors of Frequency Constants. *J.S.S.* 1908, pp. 381–397, 499–512 and 651–678. Addendum, *J.S.S.* 1909, pp. 81–90.

See text, pp. 9 *seq*. above.

52. On the Application of the Calculus of Probabilities to Statistics. *Bulletin de l'Institut International de Statistique*, 1909, tome XVIII, ler livraison, pp. 505–536.

An interesting paper showing *inter alia* the development of statistics and of Edgeworth's points of view since his earlier exposition in 1885 (No. 9 above). The penultimate paragraph contains :—" I have now touched the principal points at which the Law of Error may be applied to Statistics. Outside the domain of that law the remaining territory of Probabilities is, I think, not so fruitful of useful applications. It is a mistake to suppose that the latest developments of Probabilities have discredited the law of error. Professor Pearson has, indeed, shown that the normal law is less perfectly fulfilled than had been supposed in nature by the raw material of Statistics. . . . But at the same time he has shown how to employ the normal error-function in one or many dimensions, in a wider field and more effectively than his classical predecessors, for the purposes of calculation. . . . If curves unknown to Gauss and Laplace are required to fit the varieties of concrete Statistics, there are also required in the process of fitting new uses of Gauss' law and Laplace's Theory of Error."

53. Probability. *Encyc. Brit.* XIth Edition, 1911, Vol. XXII, pp. 376–403.

As in the case of No. 46 above, this forms a very useful summary of the original work contained in earlier papers of Edgeworth and of the work of other statisticians in the same field.

54. On the Use of the Theory of Probabilities in Statistics Relating to Society. *J.S.S.* 1912–1913, pp. 165–193. (The Presidential Address delivered before the Royal Statistical Society, December 17th, 1912.)

This is perhaps the least technical of these seventy-four papers. Among other subjects it deals with molecules as representing the statistician's ideal distribution (see No. 55), and with the practice of sampling. To a considerable extent it covers the same field as No. 52.

55. A Variant Proof of the Distribution of Velocities in a Molecular Chaos. *Phil. Mag.* 1913, Vol. XXV, pp. 106–109.

See Nos. 63, 64, 67 and 68 below.

56. A Method of Representing Statistics by Analytical Geometry. *Proceedings of the Fifth International Congress of Mathematicians* (Cambridge, August, 1912) 1913, Vol. II, pp. 427–440. Summary also printed.

57. On the Use of Analytical Geometry to Represent Certain Kinds of Statistics. *J.S.S.* 1913–1914, pp. 300–312, 415–432, 653–671, 724–749, and 838–852. Section II (pp. 724–749) circulated privately with an unpublished Appendix, pp. i–vi.

58. On the Mathematical Representation of Statistical Data. *J.S.S.* 1916, pp. 455–500; 1917, pp. 65–83, 266–288 and 411–437. Also circulated privately in a bound volume, which includes a page of corrigenda.

For Nos. 56, 57 and 58 see text, pp. 68–85 above.

59. Mathematical Representation of Statistics: A Reply. *J.S.S.* 1918, pp. 322–333.

A defence of the employment of the observed instead of the theoretical frequency in Prof. Pearson's criterion as used in the text, pp. 61 and 74 above.

60. On the Value of a Mean as Calculated from a Sample. *J.S.S.* 1918, pp. 624–632.

Dr. Isserlis' results obtained by an alternative method and justified in their general application on the principles of the law of error.

61. An Astronomer on the Law of Error. *Phil. Mag.* 1918, Vol. XXXV, pp. 422–431.

Justifies the law of error in face of certain criticisms by Professor Sampson (see No. 65 below).

62. Psychical Research and Statistical Method. *J.S.S.* 1919, pp. 222–228.

Tests certain experiments by the theory of probabilities.

63. On the Application of Probabilities to the Movement of Gas-Molecules. *Phil. Mag.* 1920, Vol. XL, pp. 249–272.

Summarised in No. 64.

64. Molecular Statistics. *J.S.S.* 1921, pp. 71–89.

See No. 68.

65. The Genesis of the Law of Error. *Phil. Mag.* 1921, Vol. XLI, pp. 148–158.

A more elaborate examination of Professor Sampson's arguments (see No. 61 above) as developed by him in 1919.

66. Entomological Statistics. *Metron,* 1920, Vol. I, No. 1, pp. 75–82.

See Nos. 40 and 50. Careful reading of these papers will show that they are not merely descriptions of a statistician's recreations, but are contributions alike to entomology and to the theory of statistics.

67. On the Application of Probabilities to the Movement of Gas-Molecules. *Phil. Mag.* 1922, Vol. XLIII, pp. 241–258.

Summarised in No. 68.

68. Molecular Statistics. *J.S.S.* 1922, pp. 479–487.

The papers Nos. 55, 63, 64, 67 and this show the application of the theory of probability to the dynamics of molecules, and are essentially of a mathematical character. It is shown that the generalised law of error applies to the physicist's conceptions (at the date of the papers at least) of molecular chaos. On the one hand, molecules give a perfect illustration of this law; on the other, the pure principles of probability are applied "to discover a law of nature" (No. 61, p. 78).

69. The Philosophy of Chance. *Mind,* 1922, Vol. XXXI, pp. 257–283.

See text, Chapter II above.

70. Statistics of Examinations. *J.S.S.* 1923, pp. 59–60.

A brief account of examination marks in India, to which the principle of No. 29 above had been applied.

71. On the Use of Medians for reducing Observations relating to several Quantities. *Phil. Mag.* 1923, Vol. XLVI, pp. 1074–1088.

Further considerations on the method proposed in (26) above, with five worked examples.

72. Untried Methods of Representing Frequency. *J.S.S.* 1924, pp. 571–594.

A plea for systematic experiments on the method of translation.

73. The Element of Probability in Index Numbers. *J.S.S.* 1925, pp. 557–575

This is a sequel to the paper, "Mr. Correa Walsh on the Calculation of Index Numbers," *J.S.S.* 1923, which was reprinted in *Papers on Political Economy*. Since the prices of commodities included are samples of prices in general, the theory of probability is needed, and is here examined with special reference to correlation between price movements.

74. Mr. Rhodes' Curve and the Method of Adjustment. *J.S.S.* 1926, pp. 129–143.

An examination and appreciation of Mr. Rhodes' suggestions for dealing with the apparent negative frequency at extremities given by the equation to the generalised law of error.

REVIEWS BY EDGEWORTH

J.S.S. 1906, p. 227. Chance: a Comparison of Facts with the Theory of Probabilities (Cohen).

1909, p. 104. Principii di Statistica Metodologica (Benini).

1913–14, p. 754. Les Mathématiques appliquées à l'Economie Politique (Zawadski).

1917, p. 546. Scope and Method of Statistics (Westergaard).

1920, p. 482. The Change in the Distribution of the National Income, 1880–1913 (Bowley).

1922, p. 107. A Treatise on Probability (Keynes).

1924, p. 103. Die Statistischen Forschungsmethoden (Czuber).

1925, p. 109. Matematica de la Mortalidal con Elementos de Probabilades (Rubio) and Introduccion a la Metodologia Estadistica (de Miguel).

1925, p. 433. A Mathematical Theory of Evolution based on the Conclusions of Dr. J. C. Willis, F.R.S. (Yule).

NOTE.—There are not improbably articles in foreign journals which are not known to the compiler of this bibliography.